Henry Anderson Bryden

Kloof and Karroo

Sport, Legend and Natural History in Cape Colony...

Henry Anderson Bryden

Kloof and Karroo
Sport, Legend and Natural History in Cape Colony...

ISBN/EAN: 9783337151454

Printed in Europe, USA, Canada, Australia, Japan

Cover: Foto ©berggeist007 / pixelio.de

More available books at **www.hansebooks.com**

KLOOF and KARROO:

SPORT, LEGEND, AND NATURAL HISTORY

IN

CAPE COLONY.

KLOOF ^{AND} KARROO:

SPORT, LEGEND, AND NATURAL HISTORY

IN

CAPE COLONY,

WITH A NOTICE OF THE GAME BIRDS, AND OF THE
PRESENT DISTRIBUTION OF THE ANTELOPES
AND LARGER GAME.

BY

H. A. BRYDEN,

Member of the South African Committee.

LONDON:
LONGMANS, GREEN AND CO.
AND NEW YORK: 15, EAST 16th STREET.
1889.

All rights reserved.

PRINTED BY
THOMAS POULTER AND SONS, LIMITED,
6, ARTHUR STREET WEST,
LONDON BRIDGE,
AND
GLOBE WORKS, RUPERT STREET, E.

To my dear Wife

I Dedicate these pages.

PREFACE.

THESE sketches and reminiscences—for they profess to be nothing more—have been written at odd times, and amid the press of other work. Yet I may hope that for those whose eyes now turn eagerly to South Africa some slight amusement, perchance some information, may be found within these pages.

To me the scenes of which these pages treat were profoundly interesting. To have wandered in the footsteps of Paterson and Sparrmann, of Le Vaillant and Barrow, of Burchell and of Campbell, and to have compared the wonderful fauna of their day with the fauna of the present; to have sojourned among the primitive up-country Boers, and heard their old-world lore and legend, and noted their quaint customs; all these were experiences of never-ending charm.

In addition to the chapters on sport and natural history, I have endeavoured to indicate in the chapters headed A Karroo Farm, The Boer of To-day, and The Future of Cape Colony, some of the present aspects of colonial life. The Cape has for too long been a neglected and forgotten Colony, far too thinly settled by our own flesh and blood. Even now, during the rush of Europeans to the Transvaal gold-fields, and beyond, the Old Colony is left behind unnoticed and unknown. This is not as it should be, for the Cape, with her millions of acres of good land now lying dormant and useless, her undoubted mineral wealth just beginning to be discovered, and her magnificent climate and scenery, surely deserves a better fate. Here, indeed, lies a country only now awaiting the presence of British capital and British blood to prove herself capable of supporting thousands upon thousands of our overcrowded population. The Cape has never yet had a fair chance. In my judgment, after many a disaster, many a bitter disappointment, the time has come when, having borne the burden and heat of the day, that chance should be given to her. Recent gold discoveries in the Colony, and the tardy appointment of a Geological Department, may tend greatly to hasten such a consummation.

In the chapter on the Present Distribution of the Antelopes and Larger Game, I have been at great pains to bring my researches down to the most recent date, and with that object I have, since I left the Colony, been in constant communication with friends and correspondents in various districts.

I have to thank Dr. Sclater, Fellow and Secretary of the Zoological Society, for permission to make use of the Society's Library, and obtain drawings (from the works of Dr. Andrew Smith and others) of some of the Cape game-birds; and I am indebted to Miss Ellen Barnes for her careful execution of these drawings. Similarly I have to thank Professor Flower and Dr. Günther, of the Natural History Museum, South Kensington, for permission to photograph certain of the antelopes—the gemsbok, the bushbuck, and the vaal rhebok—in the collection under their care.

To Mr. T. Haig-Smellie, C.E., a skilled amateur photographer, my thanks are also greatly due for his excellent photographs, from which these animals have been reproduced, and for three views of Cape scenery at pages 5, 405, and 420.

The Frontispiece faithfully reproduces a photograph of a true zebra, captured two years since in the mountains of Achter Sneeuwberg, near Graaff Reinet, and is noteworthy as being the first portrait from the life of a mature *wild* true zebra (*Equus zebra*).

I am indebted to Mr. W. MacAlister, of the firm of Wm. Dunn & Co., South African Merchants, for the use of the head of a springbok—the most perfect example of a springbok ewe I have ever seen—from which my pictures of this antelope have been photographed and reproduced.

Portions of this book have already appeared as articles in the *Field* newspaper; other portions in the *St. James's Gazette*, the *Globe*, *Home Work*, and the *Surbiton Review*. I have to thank the Editors of these papers and magazines for their courteous permission to republish here. The story in Chapter XI., A Secret of the Orange River, first appeared in a slightly condensed form in *Chambers's Journal*; and my thanks are similarly due to the Editor of that magazine for permission to reprint in these pages.

In conclusion, let me say to those sportsmen who think of exploiting the Cape, let not your motto be "slay and spare not," but rather, when possible, spare the rarer and nobler game now daily becoming more and more scarce. Remember that you may exterminate, but you can never restore. The fauna of South Africa have, during these last hundred years, after the peaceful sleep of long centuries, had a terrible awakening; it is high time they obtained some respite.

Procure necessary specimens, or meat for your camp if you will; but remember always that the game animals yet remaining to those once crowded wilds, are far more beautiful, far more useful in life, than they can ever be in death. Happily, before it is too late, preservation is coming to the rescue, and even the Dutch colonists begin tardily to acknowledge its necessity.

H. A. BRYDEN.

November,
　　1889.

CONTENTS.

CHAPTER I.
	PAGE
A Cape Cart Journey	1

CHAPTER II.
| Across Camdeboo to Naroekas Poort ... ... | 49 |

CHAPTER III.
| Klipspringer Shooting ... ... | 67 |

CHAPTER IV.
| Life on a Mountain Farm ... ... | 77 |

CHAPTER V.
| The Zebra in Cape Colony ... ... | 99 |

CHAPTER VI.
| A Race with a Kaffir ... ... | 112 |

CHAPTER VII.
| Vaal Rhebok Shooting ... ... | 125 |

CHAPTER VIII.
| A Sporting Saunter ... ... | 138 |

CHAPTER IX.
| Birds of Prey in Cape Colony ... ... | 146 |

CHAPTER X.
An Unlucky Day 156

CHAPTER XI.
A Secret of the Orange River 166

CHAPTER XII.
The Fall of the Elephant 210

CHAPTER XIII.
Springbok Shooting 220

CHAPTER XIV.
A Karroo Farm 235

CHAPTER XV.
A Morning Ambuscade on Witteberg 265

CHAPTER XVI.
The Present Distribution of the Antelopes and Larger Game of Cape Colony 280

CHAPTER XVII.
The Game Birds of Cape Colony 305

CHAPTER XVIII.
The Boer of To-day 324

CHAPTER XIX.
The Rise and Fall of Upingtonia 347

CHAPTER XX.
The Legend of Jan Prinsloo's Kloof 356

CHAPTER XXI.

THE TRUE UNICORN 386

CHAPTER XXII.

THE EXTINCTION OF THE TRUE QUAGGA (EQUUS QUAGGA) ... 393

CHAPTER XXIII.

THE FUTURE OF CAPE COLONY 404

LIST OF ILLUSTRATIONS.

1. THE TRUE ZEBRA (EQUUS ZEBRA) (*from the only photograph of a wild zebra in existence*) ... Frontispiece.
2. A CAPE CART AND MOUNTAIN SCENERY ... Facing page 5
3. VIEW NEAR VAN STADEN'S KLOOF ... ,, 17
4. A MEERKAT (PENCILLED ICHNEUMON) ,, 55
5. HEAD OF A KLIPSPRINGER RAM (OREOTRAGUS SALTATRIX) ... ,, 70
6. THE VAAL RHEBOK (PELEA CAPREOLA) ... ,, 133
7. A TROOP OF OSTRICHES ... ,, 159
8. HEAD OF A SPRINGBOK EWE (GAZELLA EUCHORE) (*Profile*) ... ,, 222
9. HEAD OF A SPRINGBOK EWE (*Full-face*) ,, 228
10. THE DAM AT RIET FONTEIN ... ,, 241
11. FLOCK OF 1,800 ANGORA KIDS... ,, 244
12. THE BUSHBUCK (TRAGELAPHUS SYLVATICUS) ,, 300
13. GAME BIRDS OF CAPE COLONY. 1. ... ,, 308
14. GAME BIRDS OF CAPE COLONY. 2. ... ,, 319
15. THE GEMSBOK OR ORYX (ORYX CAPENSIS) ... ,, 388
16. VIEW FROM BACK OF TABLE MOUNTAIN ... ,, 405
17. VIEW OF DEVIL'S PEAK, NEAR CAPE TOWN... ,, 420

KLOOF AND KARROO

SPORT, LEGEND, AND NATURAL HISTORY

IN

CAPE COLONY.

CHAPTER I.

A CAPE CART JOURNEY.

A FEW years since I travelled with two friends from Port Elizabeth up country to Graaff Reinet and thence across the Karroo Plains to a mountain farm in the Witteberg, a range west of the Zwart Ruggens district, where we afterwards sojourned. This journey was performed with a Cape cart and a pair of horses, assisted for part of the distance by a buggy and two other horses. Upon our way we encountered scenery of never-failing magnificence, execrable roads, and occasional odds and ends of shooting—foretastes of the pleasures afterwards to come. To three young fellows fresh to South Africa, the journey and its incidents were replete with untiring charms. Since that time the railway systems of the Cape have been greatly extended; but the Colony has so vast an area that even now a great portion of the travelling has to be accomplished either by Cape cart, on horseback, or by post cart or ox-waggon. The simple annals of our trip may therefore be not uninteresting to those who would care to know how progression by road is yet managed in the Old Colony. I think the majority of Cape travellers take little heed of the innumerable attractions that lie ready to the eye and hand of even the most ordinary lover of nature. The endless varieties of bird and flower and animal life in this country, possibly, have small charm for the traveller only desirous to reach his destination by the helter-skelter of the post cart, or the dull routine of the railway. But they exist in plenty by every roadside of the Colony for those who would seek them. No hedge, or wall, or fence mars the wild beauty of the landscape as you travel, save here and there near a farmstead, an ostrich camp, or cattle kraal. The terrain lies open around you, as the sky above; you halt when you like, proceed when you like, in unrestricted freedom, and you may take your meals as you do your pleasure, under the broad canopy of heaven. After a prosperous voyage from Dartmouth to Cape Town and thence to Port Elizabeth (Algoa Bay), we stayed at "the Bay," as it is colonially called, for a few days, making inquiries as to the whereabouts of Naroekas Poort, the farm whither we intended to journey. We had expected letters which should acquaint us with the means of reaching our destination, but having arrived a mail before we were expected, these letters had not come down country, and we experienced much difficulty in finding out exactly where we had to go to. We knew that Naroekas Poort was *somewhere* in the direction of Graaff Reinet, and determined, therefore, to make that town our first point.

After many inquiries we at length encountered at dinner at the hotel a young Afrikander* who was returning to Graaff Reinet, where he lived. We soon learned that he had a good Cape cart and pair of horses of his own, and that he had further bought from some hard-up colonist a buggy and another pair of horses; and after some little bargaining—for I am bound to say that the average Afrikander is an exceedingly keen hand at a deal—he agreed to take us to Graaff Reinet for £5 a-piece, this sum not to include such food and lodging as we should require on the four days' journey that lay before us. Our bargain struck, we lost no time, but arranged to start the next day, and in the meantime we were fully occupied in separating our light baggage from the heavy. The space allotted permitted each one of us to take a light portmanteau, a rug and mackintosh, as well as a rifle, shot gun, and ammunition. In addition to this we carried a long coaching-horn and a banjo, and it is due to these instruments to state that they conduced in no slight degree towards the success of our trip. With the posting-horn, until in an evil hour it perished miserably, we awoke the echoes of many a bushy kloof, many a rocky krantz, and many a stern mountain as we passed by. Merrily it trumpeted our advent to the accommodation-houses where we outspanned at night, and as blithely heralded our departure. With the banjo we charmed the inmates of many a lonely farmhouse in our wanderings, and even the stolid Boers unbent before the unwonted strains of this humble but ever-welcome instrument.

---

* An Afrikander is a person born in South Africa, of European parents.

The remainder of our heavy luggage we left behind, to be forwarded to our ultimate destination at the earliest opportunity.

All travellers by road from Graaff Reinet to Port Elizabeth, except the transport riders, as the ox-waggon carriers are here called, outspan at Uitenhage, stable their cattle there, and traverse the eighteen miles from that place to the Bay by means of the railway. From Uitenhage to Graaff Reinet by road is about 162 miles.

On a fine morning of South African spring, then, we left behind us the busy town founded by the sturdy settlers of 1820, and soon arrived at Uitenhage, after passing Zwartkops, the favourite resort of the Port Elizabeth youth, where exists a thriving boating club on the banks of the Zwartkops River. Uitenhage, which is situated higher up this river, on the slope of a hill, we found to be one of the greenest and prettiest towns in the Colony. The Winterhoek Mountains towering behind it, and here and there rising into lofty peaks many thousand feet in altitude,* form a magnificent background, and the dark verdure of the dense bush veldt, with which the neighbouring heights are clothed, is always a pleasant relief to the eye. The river, unlike most South African streams, is perennial, and the ample water-courses which lead out from a fountain in the neighbouring range and irrigate the town, the well-grown trees lining the wide streets, the pleasant, comfortable houses and spacious gardens, and the roomy, old-fashioned hotel, all combine to render this town a gracious and refreshing memory to the Cape

---

* The Cockscomb, the tallest peak of the Winterhoek, is about 7,000 feet in height.

CAPE CART AND MOUNTAIN SCENERY

traveller. A large wool-washing industry has latterly sprung into existence at Uitenhage, and the place has considerably increased in size and wealth. We soon found our way to the hotel, had some luncheon, got out the horses and traps, packed our luggage, bade good-bye to two or three friends who had come out from the Bay with us, and, with much blowing of the horn, set forth.

The Cape cart is a typical and extremely serviceable vehicle, having two wheels, a canvas tent or tilt—open front and rear, and two seats placed one behind the other, which will accommodate at a pinch four persons. All the passengers sit looking to the horses, the hinder seat forms a locker, and behind this is a swing board, as in a dog-cart, whereon can be packed a reasonable amount of luggage. To our Cape cart, which our Afrikander guide, Mr. F——, drove himself, were harnessed two of the best horses—a chestnut and a bay—I ever saw in South Africa. Well-bred from imported English sires, well-groomed—a very unusual thing—and standing higher than the average Cape horse, they pulled us gaily along throughout the whole of the rough journey, and, indeed, proved altogether too fast for the more sober pair that drew the buggy and its fortunes. The Cape buggy—also two-wheeled—sports a hood which lets down and is closed behind, and only accommodates two passengers. We found the two horses recently purchased with the buggy hardy, useful animals, but terribly low in condition, and literally covered with ticks. Evidently they had been running on the open veldt, and had not been groomed, and in consequence the ticks, which abound in most parts of the bush veldt, had acquired a terrible hold. I have stood, after we had arrived up country, for half-an-hour at a time pulling out these loathsome pests from the skins of the poor nags, each insect gorged and bloated with the blood of its unfortunate victim, and yet scarcely seeming to make any impression on their numbers. The horses drained of their strength in this way lose condition rapidly, and even die from the ravages of these bloodthirsty tormentors. The Cape horse is luckily a hardy and a willing animal, and although we travelled to Graaff Reinet over frightfully rough roads—if roads they can be called—at the rate of forty miles a day, the more generous feeding and treatment caused Sultan and Fairplay—as the old nags were called—to improve considerably in condition and in pace, even upon the journey, and afterwards when we had purchased them, they proved good and faithful servants on many a long day's trek. We fed our horses as usual on oat-hay forage, and gave them as much as they could eat. This forage was dear at the time, and cost one shilling and sixpence a small bundle, consequently it formed a heavy item in our and our Afrikander friend's bills, as we shared the expense equally.

We started at twelve o'clock. I remember the day well, for one of those unpleasant hot north-west winds, peculiar to the Cape at certain seasons, had just begun to blow, and after we had travelled a few miles we were enveloped in clouds of red dust, which found its way into our eyes, hair, and ears, and was most disagreeable. At the Cape it is the custom to outspan or unharness every two hours or thereabouts. The horses are then taken out and knee-haltered, and straightway roll in the sand or soil, after which they feed for half-an-hour, have some water occasionally, if it be handy, and the journey is then resumed. This is an ancient and invariable South African custom, and, strange as it may seem to the uninitiated, it certainly serves its purpose admirably, and the long treks, day after day, are by this means comfortably got over. Distances at the Cape are always computed by hours, six miles an hour in a cart being about the average pace. Passing Prins's Kraal, about three hours from Uitenhage, we outspanned at the farm of some young Scotch settlers. Here, while the cattle were resting, we had a splendid and most welcome bathe in a deep and rocky pool hidden in the bush veldt near the house. This was a rare luxury indeed in Cape Colony, where a plunge and short swim, such as we enjoyed, are in many places unheard of and unattainable. How we revelled in the cool and limpid depths of that pool, and how reluctantly we quitted it! While we dressed upon a flat ledge of rock a hammerkop (hammerhead) came down to the water, but swerved off on seeing us. This—the *Scopus umbretta* of Gmellen—is a curious looking bird, having plumage of a dark purplish-brown colour, a crested head, and a longish black bill. It belongs to the heron family, and feeds principally on fish and frogs. It has, further, a jackdaw-like taste for odd bits of metal, bright buttons, and such like treasures, which it hides away in its nest. From this farmhouse there was a fine view of the rolling bush veldt country lying everywhere around. Just as we had inspanned and were starting, our best pair of horses, whose heads were left for a few moments, turned short round, and before any one could reach them, bolted off down the steep hill as hard as they could pelt. As we saw them careering in the distance it seemed as if nothing could avert a catastrophe, and we abandoned all hope of seeing the cart again in any other shape than pieces. The buggy, with two of us in it, tore after the fugitives instantly, and after a chase of fifteen minutes we discovered the cart and horses round a corner of the road in charge of a Hottentot, who had most providentially managed to stop them. Giving the man a "tip," and congratulating ourselves on a lucky escape, we resumed our journey. The dust and wind were simply indescribable, and we began to think that if South African life was to be judged by this first day's sample, we should hardly enjoy it as we had expected. Happily our fears were not thus realised, and many a lovely day of perfect weather afterwards compensated us for that miserable first day's trek. Occasionally, however, the colonists are treated to these gentle reminders.

When the soft breezes of South African springtime blow gently through the deep kloofs, where the wild geranium, the heath, the iris, the lily, and the yellow acacia, and many another fragrant flower and bush bless the sunshine, and over the wild mountain tops, where the long grasses wave hither and thither at the zephyrs' bidding, or across the Karroo's broad bosom, when, after the thunder rains, the parched veldt springs, as if by magic, into a carpet of dazzling colours, then, I grant you, Africa is charming indeed, and you bless the days and extol the climate in no measured strain. But when the north wind blows fiercely, glowing hot from its passage over thousands of miles of heated plains, smothering you in clouds of hideous red dust, then indeed you feel inclined to curse the day you ever saw the country.

However, there was nothing for it but to grin and bear the infliction. At our next outspan, Blaauw Krantz (Blue Cliff), we picked up a Boer who was making for the next accommodation-house, where we proposed to stay the night. This gentleman rode with us, and peppered us incessantly with questions, which our Afrikander guide duly translated.

The ages and description of our parents, the numbers and ages of our brothers and sisters, where we came from, where we were going to, and what we were going to do, what were our possessions in land, money, and flocks—these and innumerable others were the questions asked of us. But this, as we soon learned, is an invariable Dutch custom, not intended in any offensive spirit, and the newly-arrived traveller will do well to yield good-humouredly to the rather un-English catechism. The fact is, the Cape Dutch are a simple and eminently patriarchal people, who take a keen and even a ludicrous interest in all matters pertaining to the family circle. If you would win your way instantly to their good graces, tell them you are the father of a dozen stout children. They will be enraptured, for the Boer, himself uxorious to a degree, has an intense admiration for large families and their parents. Possibly this feeling is engendered out of Nature's political economy, for South Africa is a vast country, and sparsely populated.

We were now getting well into the Great Winterhoek Mountains, and the scenery around us was simply magnificent. Abruptly turning a corner of the road we in the Cape cart came suddenly upon a very fine tiger-cat, or serval (felis serval), the boschkatte (bush-cat) of the Boers. This is one of the most beautiful of the numerous tribe of the smaller *Felidæ* that haunt Cape Colony and annoy the farmers, and is beautifully coloured in yellow and black. The cat was, however, too quick for us, and escaped into the dense wall of bush veldt that lined the road, before we had time to shoot. Towards evening, between Blaauw Krantz and Roode Wal—our outspan for the night—we sighted three or four brace of wild guinea fowl or pintados (*Numida mitrata*) running along the road in front of us. We at once pulled up, and one of our number—Bob—charged his breech-loader, got out of the cart, and softly followed them round a bend of the road, which was here composed of deep sand. Very quickly we heard the report of both barrels, and, driving on, met the gunner carrying a dead bird. The guinea fowls —always great pedestrians—would not get up, so he had had a couple of long shots at them running, and had luckily knocked one over at sixty yards. This guinea fowl is plentiful in the densely-bushed country near the Sunday and Fish Rivers; it is an excellent eating bird, bigger and handsomer than the tame species, and may sometimes be secured towards evening while roosting, as it does, out of the way of wild cats and other enemies. This was the only game we obtained this day, but the dust and wind were much against sport, and travelling, as we were, between high walls of dense prickly jungle we had not many opportunities—such as we had abundantly later on—of looking about us for a shot now and again. At dusk, after a heavy trek through deep sand, we arrived at Roode Wal, where there is an accommodation-house. Here we had hoped to get a good wash and some clear water to drink; but alas! we found that in this dry season (for drought prevailed just then) the dam containing the water supply was all but empty, and the only liquid—a villainous compound of mud and water—far too precious for bathing purposes. We had therefore to content ourselves with a dry rub—a miserable substitute at best.

However, we were shortly provided with a very fair supper and some English bottled beer, for which we paid two shillings and sixpence a bottle.

Our friend the Boer shared the meal, and I shall never forget his knife and fork play. He was a huge, slouching fellow, with a shock of dull light brown hair and heavy beard. His mouth was simply cavernous in its proportions, and his appetite what an old Cape traveller—Sir J. Alexander—would have described as a ten-pound one. When that Boer quits this earthly scene his epitaph should be something like that found in a Scotch kirkyard, which ran thus, I think:—

"Here lies Jock Gordon
Mouth almichty and teeth accordin."

Supper over and pipes lighted, we strolled out for a quarter of an hour to see that the horses were comfortable for the night, and to have a breath of cool air. The scorching wind had now fallen, and upon the heights—bush-clad as to their sides, grassy as to their crests—whereon we stood, the air was cool and pleasant, a blessed relief after our dusty and baking trek. The night was still and marvellously clear—

> "The moon
> O'er heaven's clear azure spreads her sacred light,
> When not a breath disturbs the deep serene,
> And not a cloud o'ercasts the solemn scene.
> Around her throne the vivid planets roll,
> And stars unnumbered gild the glowing pole,
> O'er the dark trees a yellower verdure shed
> And tip with silver every mountain's head."

Around us lay the silent mountains. Far below us rolled, under the silvery light of the moon, into dim distance the vast mysterious-looking sea of dense bush veldt, wherein to this day the fierce Cape buffalo and the mighty elephant find secure retreat. In this primeval shelter, wherein old "Buffel," as the Dutch call him, has couched his lowering front from time immemorial, he is likely for many decades to come to linger almost undisturbed. This region is rich in Cape history. All this wild jungle country was in bygone days the favourite lurking place of the Kaffirs in their battles and forays with the Boers and British. Amidst these bushy fastnesses the Hottentot and Kaffir rising against the Dutch farmers began in 1798, at the time when the rebellious Boers, under Van Jaarsveld, were in arms against the British, then lately possessed of the Colony. Bands of these two native races—usually implacable foes, and seldom allied in arms—sacked, burned, and devastated the frontier farms even as far as Lange Kloof, the Knysna, and the eastern edge of the Great Karroo. In much later years the Kaffirs have found sanctuary in the unknown recesses of this bush country during the numerous petty wars that have vexed the eastern frontier of the Colony. Happily,

Kaffir wars may now be considered things of the past. The density and impenetrable nature of this wall of bush veldt unless actually witnessed can hardly be realised. Forest fires will not destroy it, and its thorny recesses are, to this hour, as absolutely impervious to any but the hides of Kaffirs, buffaloes, and elephants, as they were in ages long remote. Our host told us that a few buffaloes are shot here every year; but it is dangerous work hunting them in the narrow game-paths, and the brutes are extremely cunning and dangerous. Every South African hunter is aware that this animal is classed amongst the three most dangerous of the big game, the others being the lion and leopard. As often as not in these thickets the wily buffaloes lie in wait for and chase the hunter instead of being themselves pursued; and they take no end of stopping, not to say killing.

Presently we went again within doors, and then Frank produced his banjo from its case, and we treated our Boer friend to an extemporised concert. We knew a good many old glees and part songs, and I am bound to say that the enthusiastic way in which the usually unimpassionate Dutchman applauded our humble efforts fairly touched the innermost recesses of our hearts. When we had sung "Oh, who will o'er the downs so free?" and "Oh, wert thou in the cauld blast," the climax was reached, and applauding nature could no further go. I firmly believe that our little concert that night cemented a life-long friendship between that Dutch Afrikander and the British at the Cape. After "God save the Queen," and pledging one an other in "soupjes" of Cape smoke (*i.e.*, Boer brandy), we parted for the night with mutual expressions of goodwill. There was only one bed amongst our party of four, so we tossed for positions—the lucky ones sleeping on the bed, while the two others wrapped themselves in their rugs upon the floor. The Dutchman retired to a small room at the back.

At dawn next morning (cock-crow it is usually called at the Cape) we were up, and after coffee, inspanned and started. The change from yesterday was complete. This morning, after the sun had risen in ruddy splendour over the mountains, a clear blue sky and a bright refreshing atmosphere succeeded to the burning north-west blasts. Departing with a salute from the long coaching horn, we proceeded on our way, our Boer friend riding with us for a time. The horse he bestrode was a "tripplaar," and carried him with its singular ambling pace as easily as an arm chair or a rocking horse. The triple is a sort of shuffling canter on three legs, peculiar to the Cape, and a horse that possesses it commands a higher price than its fellows. The whimsical Boers have singular notions of horse-flesh. They despise mares, and will, if they can help it, on no account ride them. What their reason for this is I am unable to state, but the fact remains. We, in this country, as often as not, prefer the mare to the gelding, as being more willing, and of higher courage.

We noticed several curious birds this morning. First, the blaauw valk (blue hawk) or chanting falcon (*Le faucon chanteur*), as it is called by Le Vaillant. This hawk has certainly a very peculiar ringing whistle, but Le Vaillant invested it with altogether fictitious attributes, pretending that it sings for an hour at a time, night and morning. It is a bold bird, and will on occasion attack and kill snakes. The hawk, sweeping about the cliffs with graceful flight, was soon far out of view. Next to appear were some of the curious black and white crows, the commonest in the Colony. This bird (*Corvus albicollis*), the ringhals kraai (ring-necked crow) of the Dutch, *Le corbivau* of old Le Vaillant, is principally of a black or bronze-black colour; the back of his neck is white, and some white feathers divide the breast and abdomen. The bill is white tipped. He is a fine big fellow and a bold high flyer, and he is by no means particular as to what he eats, even finishing off weakly lambs and goats and other unprotected prey. A near relation of this bird, the bonte kraai (pied crow) of the Dutch (*Corvus scapulatus*), is also pretty often seen, especially about the high roads, where it is perpetually hunting for garbage and offal. This bird, in addition to a white patch upon the middle of the back, has a white chest and stomach. Presently we saw and shot the red fink, sometimes called the red grenadier grosbeak (*Ploceus oryx*). These birds were just entering upon their breeding plumage, which was sufficiently striking. The male specimen shot was of a magnificent scarlet upon the upper parts, as well as upon the throat and under the tail. The rest of the body was of a rich glossy black, patches of which appeared also upon the forehead and cheeks. The wings and tail were brown. These birds were evidently flying towards the Sunday River, which ran not far away, for they are fond of damp situations.

After an hour's travelling through the mountains, which presented to our view pictures of wild and sublime magnificence, tempered, I am bound to say, however, by frightfully rough and uneven roads, we reached Paarden Poort (Horses' Pass), a well-known pass, or series of passes, that give exit through the Winterhoek range. The road here is often scarped from sheer rocks, affording very little space between the towering mountain-side and the deep and yawning precipices beneath. All this day, indeed, we toiled alternatively up heights eminently trying to horse and oxen flesh (as the bones of many a dead bullock bore witness), or down uneasy declivities almost as perplexing. After an outspan for breakfast—hard-boiled eggs, ham, and some bottled beer—we struggled onwards, until presently we began to enter upon the noorse doorn veldt (literally nurse-thorn country), consisting of bushes of a tall cactus which, like the euphorbia, exudes a milky juice on being broken. Unlike the euphorbia, however, the milk of the noorse doorn is not poisonous, and forms an excellent food for cattle, sheep and goats, which are all extremely fond of it. Care, however, has to be taken that the long spines of the plant are removed before animals are allowed to be fed upon it. The sun shone brilliantly, but the air upon these mountain ranges was clear and wonderfully exhilarating.

> "Jove! what a day; black care upon the crupper
> Nods at his post and slumbers in the sun."

No cares, indeed, had we as merrily we drove along—anon changing company by changing vehicles. The roads were rough, the joltings and bangings innumerable; but what of that?—we cared nothing for such trifles. Everything was fresh and beautiful—the sky, the air, the terrain; every bird and beast

VIEW NEAR VAN STADEN'S KLOOF.
(From a photograph by R. Harris, Port Elizabeth.)

and insect new and strange to us, though this ancient land had produced their counterparts ages upon ages back.

All this ground that we were traversing is classical. Barrow—that most accurate and trustworthy of the old Cape travellers—passed through in 1796; others have preceded and followed him, and yet, though long years have intervened since Barrow's day, the country is in most respects unchanged. It is true that the Eastern province since that time has been often devastated by Kaffir wars, and has witnessed a large influx of British colonisation —I mean that wonderfully successful State-aided Emigration Scheme of 1820. The Algoa Bay settlers have in these sixty odd years spread themselves over the country, and carved out for themselves fair homes and farms and good fortunes; and the fierce Kaffirs have been finally rolled back into their own peculiar territory beyond the Kei. Port Elizabeth, Grahamstown, and other towns have risen and prospered; but the land is so vast and still so sparsely populated that you may yet travel, often for a whole day—aye, even for days—without meeting a human being. There are in the Cape Colony thousands upon thousands of acres of rich land lying untouched that might, with irrigation and cultivation, support a good portion of that overcrowded population of which we hear so much in England at the present day. In particular, I would instance certain localities upon the Sunday River in the Eastern, and upon the Oliphant, Fish, and Zak Rivers in the Western province, which are naturally irrigated, and produce in this favoured climate astounding results in the way of corn growing—crops returning a hundred-fold and more, and crops, too, that grow twice and even thrice in the year from one sowing. Irrigation at the Cape has a great future before it, and is now rapidly coming to the front. Hitherto, to their disgrace be it said, the colonists have been in the habit of importing from £300,000 to £500,000 worth of corn each year—corn which their own country could and should have readily produced.

Beyond a ramble or two for a short time at outspans, during which we saw no quarry worth mentioning, save a fine kingfisher, the *Ceryle maxima* of Pallas, otherwise the *Alcedo afra* of Shaw, which we secured by a small periodical stream wherein a little water still lay. The largest, but not the most beautiful, of the South African *Alcedinidæ*, this bird has a crested head, plumage of a blue-gray dotted with white, stomach of a reddish colour, a white chin, and black legs and bill. Its greatest length is about eighteen inches, and its cry is loud and somewhat strident.

At nightfall we reached an accommodation-house, where we halted, having compassed forty miles of rough and mountainous travel. Away again early next morning—the third day of our trip—and on to Jansenville, a village which we reached about noon. Jansenville is a sleepy little place of about 400 inhabitants, named after an old time Dutch governor. Most of the houses belong to Dutch farmers of the Zwart Ruggens district, who only make use of them when they trek in to Nachtmaal (Sacrament) with their wives and families three or four times a year; for the remainder of the year the houses are shut up. At the inn where we outspanned we saw for the first time a Bushman, or rather woman—a diminutive, yellowish brown, and very ugly little creature, whose bare scalp showed plainly between the curious kinks of wool—each kink separate—that grew sparsely upon it. Undoubtedly I think these Bushmen were the aboriginal possessors of South Africa; but the Hottentots, Kaffirs, and Boers have driven them, as a tribe, almost completely from the Colony, and the Kalahari Desert may now be considered their true abiding place. A few broken septs, or remnants, still wander about the little known banks of the Orange River and in Griqualand West.

It has been the Boer custom to treat the Bushmen as creatures utterly undeserving of human sympathy, as wild beasts of the field, to be shot down and butchered without compunction. The Bushmen, in their turn, have undoubtedly retaliated mercilessly upon their oppressors whenever they obtained the opportunity. But I am inclined to think that these poor barbarians never had a fair chance of redemption in the old days. In more modern times, under British rule, they have proved themselves excellent servants after their kind—active, faithful, excessively sharp witted, and possessed of infinitely more fire and spirit than the Hottentots. The Bushmen in the Old Colony are now harmless enough. As hunters and trackers they are simply invaluable, and as herds excellent, and they make good and very light-weight "after-riders," or second horsemen. Between 1750 and 1760 the "Bosjesmans"—as the Boers call them—so far from raising their hands against every man, appear to have frequented the Colony openly and boldly, and to have begged and pilfered, much as did the Kaffirs subsequently. But they never appear at that time to have attempted the life of any person. Not, indeed, until the Boers began with them their invariable policy of enslaving the native races with which they came in contact did the Bushmen turn upon them. From that time until the middle of the present century war has been incessantly waged upon these miserable nomads, who in turn have shown themselves implacable and desperate foes. The deeds of blood that have been enacted by Boers and Bushmen upon each other are almost incredible in their savagery, and would fill volumes if their history could be written. The ancient Boer method appears to have been, after a foray upon their flocks, or without that pretext if in need of slaves, to raise a force of mounted men—a commando, as it is called—to track the Bushmen to their lurking-places, surround them, shoot down the men and women, and carry off the children as captives. One frontier Boer has boasted that he had assisted in the death and capture of 3,200 of these poor wretches, another of 2,700. In Barrow's time a Boer from Graaff Reinet, being asked at Cape Town if the savages were numerous or troublesome upon the road, replied "he had only shot four," with as much composure and indifference as if he had been speaking of four partridges. Barrow himself heard one of the Dutch colonists boast "of having destroyed, with his own hand, near 300 of these unfortunate wretches."

In their turn the poisoned arrows of the Bushmen not infrequently proved the death of the Dutch and their families, and they became extraordinarily expert as cattle lifters. As a race, the tiny Bushmen average only about four-and-a-half feet in height among their males and about four feet among their women. In common with all the South African races their primeval home would seem to have been in North or North-East Africa. It is probable that the three principal native races now inhabiting Southern Africa originally dwelt somewhere in the direction of Egypt or the Nile. Indications, whether in the case of the Bushmen, the Hottentots, or the great Bantu race—in which are included the Kaffirs, Zulus, Bechuanas, Basutos, and others—point irresistibly to this conclusion. The Bechuanas, even to this day, bury their dead with their faces turned in the direction of Egypt, and the huts of the Abyssinians bear singularly striking resemblance to those of the Bechuanas and Basutos. Herr Merensky, in his "Beitrage," asserts that "Many of the usages of the Kaffir tribes would seem to indicate an Egyptian origin or influence. . . . the brown people who are painted on the walls, as in battle with the Egyptians, or as prisoners, bear throughout the stamp of the Kaffirs. Weapons—the form of the shield—of ox-hide, the clothing, the type of race, are surprisingly like those of South Africa."

Barth and Schweinfurth have found very similar races to the Bushmen in North Africa. Herodotus speaks of a race in that country using a language apparently resembling their curious clicking of speech. Barrow has well pointed out their strong similarity to the Pigmys and Troglodytes of the Nile; and he further, with strong show of reason, identifies them with some of the Ethiopian nations depicted by Diodorus Siculus. The voices of these Ethiops were shrill, dissonant, and scarcely human; their language almost inarticulate; and they wore no clothing. These North African people, in defending themselves, "stuck their poisoned arrows within a fillet bound round the head, which, projecting like so many rays, formed a kind of crown. The Bosjesmans do exactly the same thing; and they place themselves in this manner for the double purpose of expeditious shooting, and of striking terror into the minds of their enemies."

As I gazed then upon the little Bushwoman squatting in the blazing sunshine at Jansenville, I could not but regard her with some feelings of curiosity and interest. There is one feature I should mention in this somewhat long digression concerning these little people; they have the most beautiful hands and feet, small and delicately formed. As a set-off, their forms are often not remarkable for beauty, and their spines are singularly curved, giving them the outline of the letter S.

Another curiosity we noticed at Jansenville. This was the most primitive billiard-room I suppose to be found in any corner of the earth. Just behind the inn was a large conical Kaffir hut. Into the top of this hut a skylight had been fitted, and a small billiard-table was placed in the centre, underneath the light. How the table got there, and whether the hut was built over it, and who were the local Roberts's and Cook's that patronised it, I know not. Perhaps the Bushman lady sometimes took up a cue! We had a knock-up for a few minutes, but the table was not a perfect one, and we soon quitted it. At Jansenville we had a delicious bathe in the Sunday River, which runs by. The river was very shallow, and did not admit of a swim, but we lay flat in the swift current upon a spit of sand, and for a short time were in a watery paradise, after our long journey.

The Sunday River here is broad, and, as is the case with all South African streams, when the rains come, exceedingly deep and rapid. It is a thousand pities that something more has not been done towards arresting the terrific torrents of water that, during the rainfall in this country, either rush headlong to the sea, or are dispersed and lost in sand and soil. The series of plateaux that rise from the littoral of the Colony into the interior in ever-increasing altitudes are no doubt primarily the cause of this waste of the all-precious element, but " damming and blasting," as someone has profanely observed, would do much to prevent this loss. A good bridge—one of the few in the Colony—spans the river here. Jansenville is not such a rising community as its position warrants. It seems that some farmers named Fourie have the right to lead away one-half of the water that flows by; in this way other farmers and villagers have been prevented from irrigating their lands, and the growth of the place has, in consequence, been greatly arrested; for without irrigation very little grain or vegetables can be produced.

After but a moderate repast of tough mutton chops, and a feed to the horses, we proceeded, our way lying for some time along the banks of the river. Not far out of the village, on the other side of the stream, grew some good-sized trees, and in their branches we noticed four or five gray monkeys gambolling. The opportunity was irresistible, so we halted. Two of us snatched up our rifles, shoved in cartridges, and took aim, the distance being some

200 yards. I am not sorry to record that the bullets missed their marks, but the amusing part of the performance was the way those monkeys streaked down the trees and vanished "*tenues in auras.*" It was astounding; lightning, "plain or greased," was as nothing to the marvellous rapidity of their descent. We were now well into the Zwart Ruggens (literally black ridges) district—so called from the dark, rough ridgy mountains that prevail here and there. Presently we crossed the river at Staples Drift—where lies a farmhouse—and soon after this point we first found ourselves upon the verge of the open Karroo Plains.

Ever since boyhood I had been acquainted, from the perusal of old books of travel, with the name of the Great Karroo, and as I now looked from a rocky eminence upon its eastern edge I could well imagine with what feelings the early Dutch pioneers, after toiling with heavy laden waggons and wearied oxen for many and many a long day's trek through the mountains, had at length, on emerging from some poort* upon its western boundary, set eyes upon the far rolling bosom of this mighty plain. Probably some old-world "Voer trekker" first entered upon this great desert—desert of mankind though not of game—between the years 1670 and 1700. What a prospect must then have been his! This huge tableland, standing 4,000 feet above sea-level, was then, and for much more than a century after, crowded with the most magnificent of fauna. The eland, the gemsbok, the hartebeest, the bontebok, wildebeest (gnu), quagga, blesbok, ostrich, and many

---

* Probably Karroo Poort was one of the first discovered passes from the westward to the Great Karroo.

of the smaller antelopes, all abounded in extravagant profusion, exactly as they had done for countless ages, undisturbed save by a few aboriginal bushmen, or wandering Hottentots, armed only with the most miserable and ineffective weapons. The lion everywhere stalked in lordly pride; the ponderous rhinoceros roamed far and wide; and the elephant wandered freely across from one grazing-ground to another. Alas, how have the mighty fallen! Most of these noble animals have been improved from the face of the earth in these regions. Truly, the Boers of South Africa have, since their landing in 1652, enjoyed a good innings in this most wonderful of game countries.

In the old days the journeys across the Karroo were, in dry seasons, attended with frightful sufferings to man and beast; and many a hardy Dutchman, many a span of stout oxen, have, when trekking from one fountain to another, gone down upon these plains to rise no more. Even at the present day, when droughts occur, sheep and goats perish in enormous numbers, and the farmers' flocks sometimes almost disappear altogether. Pringle has truly written of this parched plateau as

> " A region of drought, where no river glides,
> Nor rippling brook with osiered sides;
> Where sedgy pool, nor bubbling fount,
> Nor tree, nor cloud, nor misty mount,
> Appears to refresh the aching eye:
> But barren earth and the burning sky,
> And the blank horizon round and round,
> Spread—void of living sight or sound."

But it must not be imagined that the Karroo is a sandy worthless desert. On the contrary, its rich red soil, though baked and sun-dried by the burning sun, is capable of supporting a luxuriant vegetation, and when the rains fall the withered shrubs and plants instantly are transformed into a blazing carpet of the most brilliant flowers. Water and irrigation will some day work wonders upon these parched plains.

Neither is the Great Karroo—which is 350 miles in length, and has a breadth of seventy or eighty miles—to be imagined as one huge expanse of perfectly flat ground. On the contrary, it is studded here and there with hills and occasional isolated mountains; and periodical rivers or water-courses, dry for great part of the year, furrow its surface here and there, their meanderings marked by the stunted mimosas margining their banks. Geologists tell us that this desert, which now sustains, in addition to its indigenous springbok, mighty flocks of sheep and goats, was once a vast lake, whose waters supported the bygone saurians of a primeval period.

Water has in recent years been discovered at no great depth from the surface in many parts of the karroo, and windmills and wells are being utilised with very excellent results. Coal, too, has been found in large quantities in the neighbourhood of Aberdeen, not far from Graaff Reinet.

We now began—upon entering open country—to look about us eagerly for game, for in the mountains and bush veldt we had not had much time or opportunity to attend to such matters, in the constant struggle of difficult and uphill travelling. This day we first saw the black koorhaan—one of the magnificent bustards of the Cape. Of these wily birds we secured two brace in the heathy scrub with which the surface of the karroo is covered. Here, too, we first saw the meercat, a singular little animal belonging to the Viverrinæ. Its habits are somewhat similar to the prairie dog of North America. It lives in colonies in underground burrows, which honeycomb the veldt in places, and render galloping rather dangerous to hunters. There are two or more varieties; those seen on this occasion were little gray fellows who might be noticed sitting up on their hind legs, just outside their burrows, looking shrewdly about them. Upon the approach of danger they dive headlong into their holes with extraordinary rapidity.

Besides the koorhaans we were exceedingly fortunate in the afternoon in bagging one of the noblest birds in all South Africa—the paauw of the Dutch colonists, otherwise known as the kori bustard (the *Otis kori* of Burchell). This, as I suppose, the finest bustard in the world, formerly abounded on every open karroo of the Cape Colony, but is now much scarcer. We sighted the paauw (literally peacock—one of the absurd names of the ancient Boers) shortly after an afternoon outspan, feeding quietly some 300 or 400 yards away on the veldt not far from some mimosa bushes, in and out of which he wandered. Instantly we stopped the carts, and Bob and I, taking our rifles, made a detour, and then crept very cautiously after him, stopping as he emerged from shelter and making progress as he occasionally became hidden. At length favoured by the ground, we had arrived within sixty or seventy yards, and, taking steady aim, fired. To our intense satisfaction the huge bird fell forward, picked himself up, fluttered twenty or thirty paces, and then fell dead. Racing up to him we secured our prize, and found that by good chance both bullets had struck him—one at the base of the neck, the other through the wing and body. This was a stroke of luck indeed, and after mutual congratulations we resumed our journey with spirits greatly increased. During a stay of considerable length in Cape Colony I only had the fortune to secure one other paauw, though I saw half a dozen or more. Our captive scaled twenty-six pounds that night on being weighed, but these grand sporting birds run as heavy as thirty-five pounds and forty pounds, and in the interior even, I believe, more than fifty pounds in weight. The flesh of this bustard is delicious, and partakes of the true game flavour; and the bird is further distinguished from the rest of its family by its faculty of putting on fat as well as flesh, a quality in which most South African game birds are rather lacking.

This was a day of new sensations to us, for in addition to securing the koorhaans and the paauw, we sighted a steinbok, one of the most dainty and beautiful of the smaller South African antelopes, as well as a pair of the curious secretary birds. The steinbok which we caught sight of upon a thinly-bushed kopje (little hill), quickly disappeared, but the secretaries were for some time in view.

These extraordinary birds were descried stalking solemnly on the flats about 200 yards away. We stopped the carts, and had a good look at them as they paced steadily hither and thither, occasionally taking a short swift run in search of their prey —snakes, lizards, mice, the young of hares and game, eggs, and, indeed, almost any unconsidered trifle they may happen on. This singular bird

(*Sagittarius serpentarius*)—*Le mangeur de serpents* of Le Vaillant, the slang vreeter (snake-eater) of the Dutch—long puzzled the most learned of bird classifiers, but is now usually included, whether rightly or wrongly, amongst the *Falconidæ*.

Most people, howsoever little learned in ornithology, know the secretary bird, and it is hardly necessary to indicate at length its peculiar characteristics. It was anciently much cherished by the Boers, who retained it in a semi-domesticated state for the protection of their poultry-yards from snakes and other vermin. At the present time, however, it is only to be seen in its wild condition. At a distance the bird looks of a bluish-gray colour, but a closer inspection shows that the thighs, the longer quills, part of the tail feathers, and most of the crest are black. The crest consists of ten feathers, and it is from this curious and distinctive feature, and its fancied resemblance to a secretary's pen or quill—stuck behind the ear—that the bird's name was bestowed upon it. The secretary nests usually in the thickest parts of the thorny mimosa, and the eggs, invariably, as I am told, two in number, are of a dirty white, lightly blotched with reddish-brown, which becomes more thickly distributed at the obtuse end. I was shown an egg later in our travels, and only wish I could have secured it. Perhaps the nearest approach to the secretary bird is to be found in Jardines harrier (*Circus Jardinii*), a denizen of the grassy plains that lie between certain of the mountain ranges of Australia. This bird, like its South African prototype, does not perch upon trees, and as a rule feeds upon small snakes and frogs, in search of which it quarters the ground much as does the hen harrier of our own country.

Having sufficiently reconnoitred, and noted with our field-glass the sage and stately secretary birds—until, having somewhat too obtrusively attracted their attention, they coursed rapidly away across the plain—we resumed our trek, and once more at close of day found shelter in an accommodation-house, where, after supper, we discoursed sweet music and song to an appreciative, if scanty, audience.

Next morning, soon after dawn, and the invariable and ever-welcome coffee, we were industriously pursuing our journey over a dry karroo country, broken occasionally by slight hills and shallow valleys. In some of these undulations, among mimosa and other flowering bushes, the tall aloes towered aloft to a great height, glorious in their rich spikes of flower, some blood-red, some of a lovely pink hue. Many of the choicest wild plants of South Africa were passed in this region—the bright-red cotyledon, and many kinds of crassula and protea, and others whose names I know not, while the African briony, or wild vine, frequently clambered luxuriantly over the brush.

Amid this wealth of flower life we noticed large numbers of the showy sunbirds, sometimes also called sugar or honey birds, which are found in profusion all over this favoured land. So attractive were they that we stopped once or twice to obtain specimens—not a difficult matter—and were successful in shooting half-a-dozen of these charming birds. Amongst these we had, after a little comparison, no difficulty in identifying the *Nectarinia famosa* (*Le sucrier malachitte* of Le Vaillant), with its rich green coat, black wings and tail, bright yellow sides and lengthened tail feathers. Le Vaillant's name, "*Le sucrier*," is an apt enough one, for all these birds are passionately fond of "sweets," and obtain their sugary supplies from various flowers—and in particular the proteas—by means of their peculiarly long brush-tipped tongues. Another specimen proved to be the *Nectarinia violacea* (the *Soui manga orange* of Le Vaillant), a bird of truly beautiful colouring. The head and shoulders are wholly of a bright green, breast a rich violet, back, wings, and tail dull brownish-green, the stomach of a magnificent orange; altogether a very notable bird, even in this land of gorgeous, but too often songless, feathered prodigies. The *Nectarinia violacea* is decidedly scarcer than its congener the *Famosa*. Another sun-bird shot during this journey, a day or two earlier, was the *Nectarinia afra*, distinguished by Le Vaillant as *Le sucrier a plastron rouge*. The painting of this bird is striking. The breast, back, and head are darkish green, slightly interspersed with blue and bronze; the wings and tail are brown. At the foot of the breast there is a thin blue ring, or band, and beneath that a much broader one of red (the plastron), which nearly covers the stomach, and there are brilliant side feathers of yellow, as in *Nectarinia famosa*. We could not sufficiently admire these striking birds, and presently went on our way rejoicing. Later on they were skinned and prepared in a rough way—sufficiently, at all events, to preserve the plumage.

At mid-day, or shortly before, we outspanned for dinner or lunch—which you will—at a farmhouse standing far out upon the flat and open karroo. During the latter part of the morning we had seen a good many of the black koorhaan (*Eupodotis afra*) and heard their harsh grating call, and had shot one from the road, as well as a brace of the vaal, or gray koorhaan (another of the game-like bustards, technically known to naturalists as *Otis vigorsii*), a bird rather scarcer hereabouts than the ubiquitous black species. These gray bustards we had noticed running in the veldt not far from the road. Upon approaching them they had, as we expected, squatted, hoping, from the similarity of their colouring to the ground and karroo vegetation, to escape detection. Sidling round them in a circular movement, we approached within twenty yards of the too-confiding game, and then, as they flushed in sheer desperation, we had no difficulty in securing the brace of them.

As we outspanned, Bob and I noticed some more of the black koorhaan some little distance away upon the plain, and in the unthinking ardour of the pursuit of unaccustomed game, we cast aside the prospect of dinner at the farmhouse, and, taking some "biltong" with us, went after the birds, while the others proceeded indoors to see what they could have to eat. Walking across the heathy karroo scrub, we cautiously hastened in the direction of the nearest of the koorhaan, but, alas! just as we arrived within ninety yards, the artful rogues, which had been previously running, got up and flew off with noisy, scolding "craak," then dropped again upon the veldt. These birds will often run for considerable distances, and after approaching them for another quarter of a mile or so the same vexatious tactics were repeated. We now altered our plans, and I took a long sweep round, thus getting beyond the game. Then we approached one another, and this time the koorhaan, apparently flurried, or undecided by the new method of attack, got up well within shot, and a bird fell to each of our guns.

We turned in another direction, and after a short time a brace got up some forty yards off, of which one was secured by Bob, while the other, which lay a trifle more forward, was missed by myself. Another miniature stalk or two resulted in one more of the koorhaan being brought to bag, this time my gun proving fatal. After this we sat down upon the veldt, took out some biscuits, and prepared to dine. We had some springbok-biltong with us, but, being new to the Colony and its ways, were not quite certain how to eat it. Biltong—salted and sun-dried flesh—is a commissariat delicacy quite peculiar to the Cape. As a food it is undoubtedly extremely nourishing. It has the merit of remaining always sweet and good eating, even in the hottest weather; it occupies very little space, and in many a frontier commando and campaign of the Boers it has proved a most valuable sustenance in long marches. It is dark coloured—almost black looking—and is dried nearly to the hardness and consistency of horn, and it has to be cut with a sharp knife into very thin wafers to be eaten.

Of all this, however, we were in crass ignorance, and thrusting the curious looking stuff into our mouths, we gnawed away at it with an energy worthy of more tender viands. Our efforts were quite unavailing; do what we would, we could not separate or masticate the toughened flesh, and at last, after sucking away for some time, we finished with aching jaws our dry biscuits, picked up the game, and strolled disconsolately back to the halting-place. Here we discovered our friends inspanned and ready to start, and—to our further vexation—we further learned that they had had a capital dinner of roast mutton.

"Just our luck!" we muttered, as we hastily swallowed some coffee, and, bidding good-bye to the farmer, drove off. However, we had bagged two brace of fine bustards—for the black koorhaan is an extremely handsome game-like bird, and is excellent eating—and we soon recovered our equanimity. Indeed, upon these high table lands, especially in the spring and winter, the atmosphere is so sparkling, so bright, and so exhilarating, that the most miserable of mortals could not long resist the magic infection that attacks his spirits.

This afternoon we outspanned again at the farmhouse of an Englishman, and went inside to partake of the inevitable cup of coffee always proffered on such occasions. I think I witnessed here one of the most pathetic sights I ever saw in South Africa. The unfortunate owner of the farm—he was only, if I remember right, some forty years of age—was rapidly losing his eyesight. Day by day his eyes had grown dim and dimmer, until, when we passed, he could with difficulty perceive the blessed light of day. Here was this poor man, but a few years arrived in the Colony, his entire capital sunk in the farm, and that, I fear, not a very profitable one; a wife and five or six young children—the eldest, I think, not more than ten years old—dependent on his exertions; his nearest neighbours, God help him! Dutchmen and unfriendly, with that most indispensable, most precious possession—his eyesight—passing away from him. I shall never forget the gloom and sadness of that little household in the wilderness, the despairing husband, the anxious wife, the poor unkempt children, pursuing their little games here and there with a sort of subdued depression. They, poor mites, knew there was something wrong, though what exactly they could hardly tell. Truly a hard and bitter lot was these poor people's. Comfort under such circumstances was almost out of the question. We could only urge the man to go at once to Port Elizabeth and see an oculist, or at all events a doctor; but I fear the malady was too far gone for medical or surgical assistance to be of much avail. Often since that day have I thought of his family, and wished them well out of their heavy affliction.

At our next outspan, two hours later, we knee-haltered and turned loose the horses near a mimosa grove in a charming valley, through which the bed of a periodical river ran. One of the most interesting and astonishing sights in the bird life of this country is the extraordinary profusion of doves and pigeons to be met with in many localities. Here among the mimosa and other bushes there were many hundreds of these beautiful birds. Frank, who wanted some practice, wandered about, shooting here and there, until he had obtained some eight or ten brace of the pretty creatures, enough for a pie or two—it seemed almost a shame to shoot them for so base a purpose —when he desisted. Amongst his captures we noticed that dainty little pigeon *La tourterelle* of Le Vaillant (*Æna capensis*), distinguished by its dark red wing feathers, edged with rich brown, its glossy black head, throat, and chest, and its red bill and orange feet. Here also were specimens of another of the *Columbidæ* discovered by the indefatigable Le Vaillant, the *Columba Le Vaillantii*, commonly called by the Dutch colonists the tortel duif. Upon the neck of this bird, which is principally of a grayish brown colour, the broad black half-collar, fringed by a thinner ash-coloured band, characteristic of the turtle-doves, is at once apparent. Here, too, was the sweet laughing dove of the colonists, *La tourterelle maillée* of Le Vaillant, *Turtur senegalensis* of Linnæus, distinguished from its softcooing cousins by its singularly human-like laughing note. The noticeable features of this little beauty's plumage are the peculiar black mail-like markings upon the breast, the dark rufus-coloured shoulders, and the nearly white stomach. But wandering a little further afield to a stony kopje, we killed the first buck of our trip. This was a handsome little ram steinbok, which was shot in some straggling covert by our guide and Frank. This most elegant little antelope, notable among its fellows for its light and beautiful form, its bright red cinnamon colour, and its swiftness in the chase, is yet often very easy to obtain; for it has a most stupid habit, especially in bushy cover, of stopping to stare at its disturbers. The little ram shot on this occasion had almost waited until the gunners were upon it, and had then been easily shot at close quarters. Having fastened up the steinbok upon the Cape cart, we drove steadily on till nightfall.

Next morning we were early upon the road, and after half a day's trek across the brown karroo, which here stretched everywhere around, bounded only in the distance by the great Sneeuwberg range, we began to approach the first goal of our journey, Graaff Reinet. Shortly before reaching the bend of the Sunday River, which almost completely girdles that town, and has to be crossed at a dangerous drift, we passed a flock of ostriches, driven by a Kaffir boy, marching along in stately procession. These birds are frequently driven about—if at all difficult to manage—with leather thongs passed round the chest and under the wings; they are then guided with little trouble by two men. We had now to effect the rough and difficult passage of the Sunday River, whose bed, now that the water was low, showed plainly the huge boulders and deep holes that beset travellers crossing these drifts.

We could at length behold the quaint and delightful town that lay before us basking in the clear spring sunshine. Lapped in peaceful repose, clad in verdure, backed by the towering masses of mighty Sneeuwberg, beneath which it nestles, and having an abundant water supply from the river as it runs by, well has it been christened by travellers of old "the gem of the desert." For to the wanderers wearied by drouthy, baking, summer treks across the burning karroos that everywhere encircle it, this, to my mind, the most charming and characteristic old Dutch town in all South Africa, must have seemed a veritable paradise.

At the drift (ford) of the river, before getting into the town, the banks are extremely steep, and the descent is a long and protracted one. Bob, who was driving the buggy, let his horses have their head a little too freely in making the descent. They quickly degenerated from a trot to a gallop, and clattered down the hill, through the boulder-strewn river, and half-way up the other bank in a rather alarming manner, so much so that we were both surprised at having been able to keep our seats in the process. On these occasions the traveller's feelings are very much those of a shuttlecock—if shuttlecocks could feel—for they are almost absolutely helpless.

Just at this hill a heavy transport waggon, with a span of fourteen fine oxen, took the drift from the Graaff Reinet side. We had often met and admired on our journeyings the old Cape waggons, with their long teams of oxen, more even, perhaps, than the teams (almost equally useful in their way) of mules and donkeys that ply with lighter waggons upon the colonial roads. But this particular waggon was so well equipped, and the oxen were so comely and in such good fettle as to arrest our attention. I suppose there was a load of 7,000lbs. weight at the very least upon the waggon, but the great oxen were so stout, and faced the long, steep hill so gallantly, amid loud cracks of whip and lusty cries of the little Hottentot that drove them, that we paused to watch the stirring spectacle. It was a fine thing, albeit, in South Africa, a sufficiently common-place matter. The Cape oxen are in truth very different beasts to those we are accustomed to see in England— especially the high-shouldered Fatherlanders. Tall, gaunt, and bony, they attain a mature age, and are not slaughtered, as in this country, before they are full-grown. Their strength is immense, and when they like, and are in a mind to bend to the yokes together, they can pull indeed. In the history of South Africa the Cape oxen have played an all-important part. Hardy and rugged as their Dutch masters, their sturdy efforts have conquered for the early pioneers the almost insuperable difficulties of this wild country. Rugged mountains, dangerous kloofs and poorts, rushing rivers, parched and waterless deserts, all have succumbed before their untiring and ill-requited exertions.

It is, indeed, not too much to say that but for the patient ox the Boers would never have penetrated, as they have done, to the uttermost recesses of Southern Africa. But the ox has his bad qualities. He is at times obstinate, ill-tempered, and treacherous, and with increasing age and much trekking he becomes " as artful as a waggon-load of monkeys." The Boers are extremely proud of their cattle, though they often sadly ill-use them. It is the thing with them to possess a span of oxen of even size and colour, and yellow oxen are oftentimes the most admired.

Unless you know them well, however, the Cape oxen are " kittle cattle " to deal with. I remember well the first time I closely inspected a span, which had been driven by a wandering fruit Boer up to the outspan at Naroekas Poort, where I sojourned.

As the great gaunt fellows stood lazily looking about them, I went up to one and patted him on the ribs. Instantly he let out in front of him with his near hind leg, caught me full in the centre of the thigh, and sent me flying to earth. I really thought for a moment my leg was broken, as, indeed, it might have been, and it was more than a week before I recovered the use of the limb. After that salutary lesson I have always been more careful in approaching strange oxen.

We were quickly into Graaff Reinet, and outspanned at the comfortable, old-fashioned hotel in the Market-square, where we enjoyed the unwonted luxury of a bath and a good dinner of rump-steaks and fried eggs—a standing dish in South Africa—peach pie, and some excellent bottled beer. Then we sallied forth to inspect the town.

Graaff Reinet, which takes its name from the surnames of the then Dutch Governor, Van der Graaff, and his wife's family—Reinet or Reynet—was founded in 1784, and for very many years, until Grahamstown and Port Elizabeth rose into eminence, was the principal, and, indeed, the only large town in the eastern part of the Colony. The streets are broad, well-kept, and laid out at right angles in the prim fashion of the Dutch; they are further supplied with broad channels of clear water led out from the Sunday River, and are beautified by rows of trees—oak, syringa, orange and others, which impart a delightful aspect and give grateful shade. The houses are mostly old-fashioned, roomy, white messuages, quaintly peaked and gabled after the old Dutch manner, and lit up by sun-shutters and windows of vivid green, and often covered with deep and overhanging thatches. The vineyards, orchards, and ample gardens add greatly to the beauty of the town. The interiors are wonderfully good. The rooms in these old houses are exceedingly lofty and handsome, and of quite refreshing coolness. The inhabitants number about 5,000, and are well provided with churches, banks, shops, and hotels. Indeed, I think we found the accommodation here better than anywhere else in the Colony, with, perhaps, one exception. In old days, when Graaff

Reinet was the capital of the Eastern province, the quiet ebb and flow of its existence were quickened and diversified by the advent of the trading waggons from the far interior, loaded up with ivory, ostrich feathers, hides, and other produce, and the market-place was then a brisk and strikingly picturesque emporium. In those days it must have been a sight indeed to see, as I have seen elsewhere, the great waggons outspanned, their tilts worn and discoloured from long journeying, or patched with green hide, which had hardened in the sun to the consistency of iron, and the worn oxen released from their long and fatiguing trek from distant Transvaal or still more distant Bechuanaland. Many a traveller has started thence for the far interior. Captain Cornwallis Harris, that most accomplished of South African hunters and naturalists, whose magnificent illustrated works will remain monuments for all time of the splendid fauna of the country, set out from here in 1837 on his expedition to the then unknown and unexplored Transvaal, at that time held by the fierce Moselikatze and his renegade Zulus, who afterwards formed the present Matabele nation. Gordon Cumming passed through more than once on his way down country from those wonderful shooting trips of his, his waggons loaded up with many a heavy tusk of ivory and many a goodly spoil of the chase. Barrow, Burchell, Campbell, and a host of others, all have rested here.

We much enjoyed strolling about the quaint old town, and were introduced by our Afrikander friend and guide to his fine old Dutch house and to his sister, a young lady to whose excellent singing and playing we were several times charmed listeners.

A notable feature in these old houses is the delightful "stoep" or raised verandah, where you may sit or recline sheltered from the sun, and enjoy the luxuries of tobacco and coffee in a calm and cool repose. No better "loafing" or "lazing" place was ever designed than the good old Dutch "stoep."

On reaching the hotel again we found the town band preparing to practice for a forthcoming dance. Listening to them and smoking were several people, amongst them Mr. Pieter Maynier, familiarly called by Graaff Reinetters, "Oom Piet" (Oom, or uncle, being a term of affection in South Africa), a well-known and respected gentleman, descended from one of the old Dutch families. Fired by the accustomed sound of dance music, we young fellows began to waltz with one another. Presently "Oom Piet" joined in, as well as others, and I am bound to say that the old gentleman's *trois temps* was surprisingly good—quite equal, indeed, to that of many of our fleet-footed English girls. Finding this slim, active old gentleman such a treasure in the terpsichorean art, I wisely, if greedily, stuck to him as a partner, and enjoyed a really good dance. It was a strange freak of destiny to find oneself dancing with the descendant of that old Dutch Landdrost Maynier, of whom I had read in Barrow's and other old books of travel. This Landdrost Maynier was a character of note in the troubled times at the end of the last century. When the British took possession of the Colony, he was appointed Resident Commissioner of Graaff Reinet, then one of the four divisions of the Colony, and rendered very important services to the English in the Hottentot Rebellion of 1798, and in the Boer rising under Van Jaarsfeld and Prinsloo about the same period. At length our impromptu dance was ended, and we finished up the remainder of the evening—extended to a late hour—with some very enjoyable singing.

Next morning we were up betimes, looking at new horses, inquiring our route to Naroekas Poort, and attending to various other matters that concerned us. Of course we went to see the morning market, a great feature in every South African town, where, from seven to eight o'clock in summer, or eight to nine o'clock in winter, all kinds of produce are put up and sold by auction to the highest bidder. It is a singular and striking scene. In the market-place are to be found wool, angora hair (*i.e.*, mohair), corn, skins, feathers, vegetables, fruits, butter, ducks, poultry, meat, tobacco, honey and other things, often in great plenty and confusion. Round about stand groups of people, interested and eager, or idle and inert, as the case may be, of mixed nationalities, British, Boers, Jews,—Hamburg and otherwise,—Germans, Kaffirs, Hottentots, Basutos, Bushmen, *cum multis aliis*. Here is a list of morning market prices during the current year, from which it will appear that the actual cost of living at the Cape is by no means an extravagant item :—

GRAAFF REINET MORNING MARKET.

Almonds, 200 lbs., per lb., 1½d. to 2d.
Apricots.
Angora Hair, 3,000 lbs., per lb., 4½d. to 9d.
Barley, 40 lbs., per 155 lbs., 5s. to 5s. 6d.
Beans, 200 lbs., per 100 lbs., 12s. 6d. to 11s.
Beef, 2,000 lbs., per lb., 1d. to 2½d.
Brandy, per 16 gallons, 30s. to 35s.
Buckskins, 1,200, 1s. 3d. to 3s. 1d.

Butter, 200 lbs., per lb., 4½d. to 9d.
Chaff, 8,000 lbs., per 100 lbs., 9d. to 1s. 3d.
Ducks, 40, 1s. 3d. to 1s. 9d.
Firewood, 40, per load, 15s. to 24s.
Fowls, 40, each, 9d. to 1s. 2d.
Meal, 40, per 200 lbs., 12s. to 16s.
Mealies, 40, per 200 lbs., 5s. to 7s.
Oat-hay, 8,000 lbs., per 100 lbs., 3s. 6d. to 5s. 6d.
Oranges.
Onions.
Ostrich Feathers, 200 lbs., per lb., 10s. to £9 10s.
Peas, 200 lbs., per 100 lbs., 6s. to 8s.
Peaches (unpeeled), 200 lbs., per lb., 1½d. to 2d.
   Do. (peeled), 200 lbs., per lb., 4d. to 6d.
Pigs.
Potatoes, 1,000 lbs., per 100 lbs., 3s. to 8s.
Pumpkins.
Quinces (dried).
Raisins, per lb.
Sweet Potatoes, 10, per bucket, 9d. to 1s. 6d.
Sheepskins, 1,200, 1s. to 3s.
Soap, 100 lbs., per lb., 3d. to 4d.
Stinkwood.
Tobacco, 2,000 lbs., per lb., 2d. to 3d.
Tallow, 200 lbs., per lb., 1½d. to 2d.
Walnuts.
Wool (unwashed), 6,000 lbs., per lb., 4d. to 5¼d.
Yellowwood, per foot, 3d. to 3¼d.
Wheat, per 200 lbs., 9s. to 11s.

A. B., Marketmaster.

At the time I write of, some items, such as ostrich feathers or mohair, were higher in price, but as a whole, this list fairly illustrates prices in recent years. Various calls and matters of business occupied our first day. That evening a great musical event took place at the Town Hall, for no less a personage than Madame Anna Bishop— a lady well known in Europe in the forties as a *prima donna* and concert singer—was to appear. Of course we attended on such an occasion. The hall was crowded to the uttermost. Madame Anna Bishop appeared, and, although of course her voice had long since lost its pristine freshness, her execution was faultless. The Graaff Reinetters were enraptured, and applauded with a quite astonishing vigour. Madame Bishop was assisted by a Mr. Lascelles, who played her accompaniments, and himself sang " The Wolf," " Rocked in the cradle of the deep," and other good songs at intervals, with great effect. It was strange to hear this lady —who forty years before had, with her singing, delighted the contemporaries, I might almost say, of my grandmother—in this far-away town in the South African wilderness. I believe, at the time I heard her, Madame Bishop was at least sixty-three or sixty-four years old, yet (such are the inscrutable ways of woman) she appeared upon the stage apparently not more than middle-aged. Her dresses (for we heard her a second night) and jewels were unexceptionable, and we marvelled how she had preserved the splendour of her apparel thus undimmed, for she had but just completed a tour through the Transvaal and Orange Free State, where travelling is even more difficult, and roads are far rougher than in the Old Colony. I suppose this indefatigable lady, who in her youth had charmed every capital in Europe with her magnificent singing, had seen more of the world and sung in stranger places than any other person known to fame. Besides securing a European reputation in its day unrivalled, she had sung in Tartary (at Kasan the capital), in all the known cities and towns of North and South America — crossing the Andes in this last country — and in the Australias. I met this wonderful old lady afterwards at the Phœnix Hotel at Port Elizabeth, at which town she was giving concerts. Since that time, I regret to say, she has gone over to the great majority.

We spent three very pleasant days in Graaff Reinet, visiting, among other places, the well-known Valley of Desolation, a ridge of rocks from which huge basaltic pillars thrust themselves abruptly skywards to heights of 300 and 400 feet, forming a very striking spectacle. After inspecting a good many horses and traps, we arranged with our late guide for the purchase from him of his Cape cart and the two older horses which had drawn the buggy. For the Cape cart, which was a very good one and nearly new, £45 was paid, while the horses cost £23 each. Experience afterwards taught us that we had paid quite sufficiently for these appurtenances of travel, the nags in particular not being worth the money. However, in such transactions in every land, and especially with Afrikanders, the motto is *caveat emptor;* and, as we had no adviser at hand, we might have done worse, for both cart and horses served us well and faithfully long afterwards. We were almost sorry to leave Graaff Reinet; the rest and the quiet, old world calm of the charming little town is very pleasant, and for those in search of change and absolute repose I can recommend no more suitable "outspan" for a time. The place had a curious charm for me. As I stood in front of the hotel, looking on to the Market-square and the old-fashioned houses around, I could see distinctly before my mind's eye many a bygone incident in the history of the stormy past. I could picture the arrival in the town, in 1796, of Mr.

Barrow (afterwards Sir John Barrow, of the Admiralty), most diligent and observant of officials, at that time private secretary to Earl Macartney, first British Governor of the Colony ; then the approach of the turbulent Boers—most of them the lawless freebooters of Bruintjes Hoogte, leather-clad hunters, armed with immensely long flint "roers"—in Van Jaarsfeld's insurrection two years later, when they lay encamped before the town for a month, at that very Sunday River drift where we had crossed, threatening to hang the Landdrost and garrison. Then the forced march from Algoa Bay, upon the road we ourselves had followed, of gallant General Vandeleur, with his clean-shaven, queued, stiff-skirted troops, and the retreat of the Boers before them. Then the voluntary and courageous expedition of Landdrost Maynier, unarmed, into the Sunday River bush country to pacify the insurgent Hottentots, at that time united in arms with the Kaffirs, and wreaking vengeance upon their ancient oppressors, the Dutch. Still later the risings of Bezuidenhout and Hendrik Prinsloo, and that fearful scene of execution of Boer rebels at Slaghters Nek. Brave old Sir Andries Stockenstroom, the fiery veteran, Sir Harry Smith, and many another notable character, passed before me. All these, and a hundred other moving incidents and striking figures of bygone times, came thronging irresistibly to my imagination. The internal history of the Cape Colony, of which so little is known even to the colonists themselves, would, I am convinced, if published, present one of the most striking, most picturesque, and most exciting themes ever placed before the reading public.

We had learned from Mr. Piet Maynier, who was acquainted with our future host and hostess, the exact locality of our destination, Naroekas Poort; and, further, that we had come some distance out of our way, and should have to partly retrace our steps, for two days or more, in an oblique south-easterly direction, across Camdeboo. We had also very fortunately chanced upon a young English farmer, travelling for a good part of this journey by the same road. On the fourth morning, therefore, of our stay in Graaff Reinet, once more we collected our impedimenta, blithely blew our horn, and drove out of the town.

Chapter II.
ACROSS CAMDEBOO TO NAROEKAS POORT.

WE were now to cross the ancient Hottentot region of Camdeboo, which, forming the north-east corner of the Great Karroo, is so often referred to by old travellers, very frequently in terms of anything but pleasure; for droughts frequently prevail here, as in other parts of the Karroo, and for long the Bushmen swarmed here and proved exceedingly troublesome.

The karroo is, no doubt, for a great part of the year exceedingly parched, and, to the outward eye, arid and barren; but old travellers have somewhat exaggerated its terrors. At the present day, the Karroo graziers feed enormous flocks of sheep and goats upon its dry but nutritious and aromatic herbage, and, as I have before pointed out, the rains completely transform and revivify its vegetation. Cattle and horses, too, do exceedingly well upon the Karroo pasture. Old travellers too frequently, I think, pointed to the dark side of the picture, and omitted the brighter aspect. Here is Le Vaillant's summary of this plateau land:—"We now entered the dry and sterile plains of Carouw. The earth, or rather sand, only bounded one way by the horizon, is covered with rank unwholesome weeds, and presents the mind with an idea of famine and desolation; here and there a few scattered spots of grass spring up reluctantly, offering our cattle a scanty sustenance; the other sides are bounded by steep and craggy rocks, whose appearance is not less dreary." This was the impression, doubtless, of a traveller wearied and disgusted by a fatiguing trek during great drought; for the karroo is mainly composed of rich soil, not sand, and, as I have pointed out, the vegetation by no means consists of "rank, unwholesome weeds." Still, I must admit, that even at the present day karroo travelling in summer and times of drought is exceedingly trying. We jogged merrily along, under clear unclouded skies, till we reached Hoog Kraal, on the Sunday River, where we made our first outspan. Here we breakfasted on ham, cold eggs, and cold tea, which we had taken the precaution to bring with us in Silver and Co.'s excellent vulcanite water-bottles. These bottles, covered with felt, which, when damped, renders the liquid within wonderfully cool even on the hottest day, we found simply invaluable in our hunting expeditions later on. As a proof of their excellence, I may mention that I have several times sent them out to colonial friends at their particular request, since my return to England. It is so difficult to get hold of a good water-bottle for a hot climate, that I may be excused for referring so particularly to these most excellent inventions. At the water here we shot a brace of the pretty so-called Namaqua partridge (*Pterocles tachypetes*), the common sand-grouse of the Colony, which at early morning and evening may be seen at streams and fountains in very large numbers. Although a true member of the sub-family, *Pteroclinæ*, or sand-grouse, its shrill cry and flight very much resembles the plover's. Yellowish-brown in colour, it is, like the rest of its genus, distinguished by the chest bands, which in this instance are first white, then reddish-brown. Here, too, we first saw that fine eagle, the Senegal eagle (*Aquila senegalla*), commonly known in the Colony by its Boer name, coo vogel. I think this rapacious gentleman must, like ourselves, have been attracted by the Namaqua partridges, for he appeared above us almost simultaneously with the reports of our guns. Probably he had been hanging about the vicinity for some time, waiting for a fair chance of a stoop. These birds, which in general appearance are of a reddish-brown colour, are abundant upon the karroo, and are the frequent attendants of shooting-parties, when they not infrequently stoop and secure a wounded bird under the very eyes of the gunner. On this occasion the eagle had no chance of thus snatching our game from us, and after wheeling about for a short time well out of range, finally betook himself off with bold and rapid flight.

We had expected to see springbok upon the Camdeboo Plains, and we were not disappointed. Towards eleven o'clock our guide pointed out to us some whitish-looking specks upon the veldt, some distance in our front. After proceeding a few hundred yards farther, we could just distinguish the forms of the antelopes, but they were extremely wild, and bounding like india-rubber balls into the air for a few strides, stretched themselves upon the karroo, *ventre à terre*, and were quickly far out of sight. There were about two hundred of them, but I suppose they had been recently hunted, and we never had even a chance of a shot upon this our first acquaintance with them. On subsequent occasions, however, we interviewed the springboks with much more satisfaction to ourselves if not to them. This day we saw a good many koorhaan of the black and gray varieties, but as we were anxious to get on to our destination, we left them undisturbed.

It was a glorious day, and we greatly enjoyed the magnificent prospect around us. People speak of the monotony of these plains, as if nothing ever occurred to break the far-extending view. It is true in one direction the terrain rolled without a break to the horizon, but on our right flank the bold range of the Camdeboo Mountains heaved from the plain in magnificent grandeur. Nothing is more striking to the new-comer at the Cape, than the marvellous transparency of the atmosphere. Mountains that appear about five miles distant, to any one freshly arrived from the comparatively thick atmosphere of Great Britain, are, more often than not, thirty or forty miles away. For this same reason, rifle shooting at long ranges is at first extremely disappointing; the buck that appear so close, are often four or five hundred yards distant, and your bullets strike up the dust far on the hither side of them. I think the aspect of these mountains that border the Karroo are even more striking and more beautiful to witness than those seen in less flat regions. Their magnificence is intensified by the sharp contrast, and whether you gaze upon them under the purple blush of sunrise, or at noon when upon their brown sides, seared and torn by wear and weather of centuries, or by volcanic action, or both, you may pick out their every cleft and furrow, even though you stand twenty or thirty miles away, or watch them in the far-away distance fading into a dim cloud of blue upon the horizon; or whether you see them bathed in the soft rosy hue of evening, you are uncertain under which of these changing and ever-glorious aspects they are most beautiful. During the whole of this day's trek—over forty miles—such is the translucency and rarefaction of this perfect atmosphere, the most prominent of the Camdeboo range, a bold mountain standing out from its fellows, seemed almost as near to us at close of day as when we passed it in the early morning. One more note upon the splendid air of these Cape uplands and I have done. It is probable that few people at home are at all aware of the capabilities of this climate as a means of restoring invalids, and especially those having consumptive tendencies. Here is a summary of Karroo atmosphere by Dr. Hermann Weber. Its notable qualities, he tells us, are as follows :—

1. Purity: comparative absence of floating matter.
2. Dryness of air and soil.
3. Coolness or coldness of air temperature, and great warmth of sun temperature.
4. Rarefaction.
5. Intensity of light.
6. Stillness of air in winter.
7. A large amount of ozone.

To this list I would add:

8. Night-time nearly always cool—a very important matter, in my humble opinion, where invalids are concerned. For myself, although by no means in need of the recuperative qualities of the Cape climate, I can say that I never felt better at any time of my life, than when enjoying the dry, pure, and intensely exhilarating atmosphere of these Karroo regions.

We saw to-day a few *black* crows, a rather unusual sight in this part of the Colony, where these gentry are almost invariably pied, black and white.

This was the korenland kraai (cornland crow) of the Dutch, *Corvus segetum* of Temminck. In size it is identical with the black and white varieties, the bonte kraai (spotted crow) and ringhals kraai (ring-neck crow), but it is more commonly found in the Western province, where the agricultural farms are more plentiful.

There is to be found upon the karroo, in addition to the numerous game birds (always interesting to the sportsman), a store of other birds, but little noticed by the average passer-by, yet extremely interesting to those who will take the trouble to look for them. We had not much time, except at outspans, to notice the smaller avi-fauna; but later on, when staying upon these plains, we were much interested in watching the wealth of feathered life around. At one of our outspans this day, I shot in a clump of doorn-boom (literally thorn-tree) or mimosa bush, a bird of the woodpecker family, often found in these regions. This was the *Laimodon unidentatus* of the well-known traveller and naturalist, Lichtenstein. Of a smallish size; in colour, upon its top, black, streaked with yellow, and having a rich crimson forehead; a stripe alternately yellow and white extending over the eye to the back of the head, and another white line down the side of the neck; a black gorget, a greyish stomach, and armed with a strong black beak—this bird is always seen in pairs. It is pretty readily capable of domestication, and is often seen in the bushes or small trees near Karroo farmhouses. Its cry consists of three notes, something like "Poo-poo-poo." Doves, especially *Æna capensis*, and many other birds abounded in this mimosa bush, and the schaapwachter (shepherd) of the Boers, a

A MEERKAT (Pencilled Ichneumon).

saxicola, and one of the most charming songsters of the Colony, was often seen and heard. Meercats were abundant, and a jackal once or twice might be seen prowling in the distance.

The little grey meercat (literally moorcat), commonly found on the Karroo, is, nowadays occasionally seen about farmhouses, quite as tame as an ordinary cat. The picture accompanying will best demonstrate the peculiarities of shape of this singular little animal. One of the Viverridæ, and known to naturalists as the Pencilled Ichneumon, this meercat, as I have previously mentioned, lives in colonies in burrows on the karroo. Its singular forearms are evidently designed for burrowing, and are really two sinewy little spades of flesh and blood. I first saw one tame at Riet Fontein, where we afterwards stayed. The little animal was made a great pet of, and amongst the members of the household it displayed the greatest affection. Towards strangers it was not so friendly, and with them it would sometimes display, and even use its sharp little teeth. I remember a very ludicrous instance of this hostility occurring. Vrouw Stols, wife of a neighbouring Boer, came up to the house for something; while she stood talking at the door, Kitty, the meercat, who had silently approached her from behind, bit her sharply on the heel just below the ankle. The good Vrouw, like all Boer ladies, wore low veldt-schoons and no stockings, and the sharpness and suddenness of the attack made her leap into the air screaming vigorously. Then she fled incontinently homewards, only to be brought back again. When an explanation took place, Kitty was scolded, and the stout Boer lady at length mollified. When stroked or scratched, and made a fuss with, Kitty would utter a curious low, throaty cry, as indicative of pleasure, as is the purr of a cat. She was desperately afraid of vultures and other birds of prey. Sometimes the children, whom she often followed out of doors, would imitate the cry of an aas-vogel (vulture). Instantly Kitty would run, in the greatest alarm, to the nearest person for shelter from her aerial foe, which she evidently imagined to be meditating a stoop at her. I believe my friends at Riet Fontein were among the first to tame and domesticate the meercat. Their example has been followed by others, and this entertaining little creature now bids fair to become a universal pet among Cape colonists.

With a few rough exceptions upon rocky undulations in the plains, our journey progressed smoothly enough until nightfall, when we reached the hospitable shelter of a large Karroo farmstead, owned by two brothers—Englishmen. Here we first witnessed one of the most remarkable and striking sights to be seen at the Cape—the return to kraal in the evening of the vast flocks of sheep and goats that are daily depastured upon these mighty plains. Near the house lay a great dam, and at the accessible end of this dam the flocks, to the number of many thousands, in charge of the Kaffir herd-boys who tend them during the hot hours of daylight—moving as they move, and keeping watch, just as David did, for wild animals such as jackals and the like—come to quench their thirst after their long and waterless hours of pasture. It is a most singular and impressive sight to English eyes: then the refreshed animals pour quietly into the kraals or enclosures for the night. After a cheery evening with our guide and hospitable entertainers, we betook ourselves to rest, and, as bed accommodation was scanty, I slept with Bob on some empty sacks upon the floor, having, as usual, lost the toss for beds. As is often the case in spring upon these elevated plains, there was a severe frost, and the cold was intense, although we were enveloped in our thick bush rugs; we were, therefore, not sorry when morning and the welcome sunshine again appeared.

The bright morning star, familiar to every South African hunter and traveller, had not long shot swiftly above the horizon, before we awoke cold and hungry, but refreshed withal from our slumbers. It may surprise many in this country to know that nights are so cold and even exceedingly frosty in the spring and winter of the Cape. Such is, however, the case; and upon the huge plateau of the Great Karroo (3,000 feet and more above sea-level), the second of the series of terraces which rise in increasing altitudes from the Indian and Atlantic Oceans to the Orange River and beyond, very severe frosts are experienced, and even the springbok and other game rise stiff with cold at these seasons. We had a long day's trek before us, and having partaken of coffee we inspanned our two good horses, and were ready to start.

Meanwhile, one of our hosts was superintending the unkraaling and counting out of the seemingly countless flocks of sheep, which defiled in charge of Kaffir herds, drinking as they went at the dam hard by, for their day's pasture far out on the open and apparently illimitable plains. These plains are sparsely covered with a growth of low scrub, in appearance not altogether unlike Scotch heather, and, except when the rains fall, their expanse presents a brown and somewhat barren appearance. When the rains come, the karroo, having naturally a fertile soil, starts at the touch of moisture into an immense and beautifully variegated ocean of flowers, which, in turn, shortly disappears under the hot African sun. Notwithstanding drought, the sheep, goats, and ostriches of the Cape farmer flourish exceedingly upon the parched herbage; and were it not for the ever-present difficulty and danger of the lack of water, pastoral farming would be a calling fairly free from care. Quite recently, wells and windmills have been employed on these plains with successful results, and there are people who prophecy that one day the parched face of the Great Karroo will smile under yellow crops of corn. Undoubtedly irrigation and tree-planting have a brilliant future before them in the Cape Colony, which in years to come will be a great grain-producing country. The kraals are here composed of high walls of solid bricks, cut out of the enclosures themselves, from the dung of the sheep, dried by the sun, and hardened by the constant trampling of thousands of ovine feet, almost to the consistency of kiln-dried bricks. This substance also, under the Boer name of "mist," constitutes with a little wood the principal firing of the Karroo farmer. Bidding good-bye to our hospitable entertainers, we entered upon the rough track, called in these regions a road, and with the earliest gleams of morning light pursued our way. Almost immediately, amidst a blaze of ruddy gold, the sun-god rose in radiant magnificence sheer upon the plains, which to the east stretched without a break to the horizon. The range of mountains on our right flank, which we quitted yesterday at mid-day, but which, although forty miles away, appeared in the rarefied atmosphere of this climate scarcely a fourth of that distance from us, was tinted with hues of the tenderest rose, fading here to warm brown, and there to a soft purple. The air was bracing and exhilarating to a degree, and for a time our coats and waistcoats were still welcome. Presently, as the warmth increased, we doffed these garments, rolled up our sleeves, pulled forward our broad felt hats, and were prepared for a hot day's travel. We had guns and rifles ready, and after awhile, seeing a koorhaan not far from the road, one of us jumped out and stalked quietly up to the spot where it was last seen, while we others in the cart moved quietly along. The koorhaan is a wary and indeed a vexatious bird, but on this occasion he was careless, and just as he rose five and thirty yards away, with his harsh grating cry, he was brought to earth again and quickly bagged. The tedium of Cape travel is happily often relieved by a little welcome sport on the way; sometimes a koorhaan, sometimes a brace of sand-grouse (Namaqua partridge), or francolin, or guinea-fowl; occasionally a springbok.

Quickly we resumed our journey, nor drew rein again till we outspanned for breakfast, two-and-a-half hours from the farmhouse we had quitted.

The horses were unharnessed and knee-haltered, when, as is their invariable custom, they had a roll in the dust, and then proceeded to graze quietly upon the veldt and the oat-hay forage with which we provided them. There was no water until four hours farther on, so they had perforce to do without. Meanwhile, we breakfasted upon sardines, springbok-biltong, from which we cut thin wafers, and biscuits, moistened with a little hollands (out here called "square face") and water. Pipes were lighted, and in less than half an hour we inspanned our cattle and started again; upon the flat Karroo roads we slipped merrily along, though here and there spruits, stones, and dry river-beds shook us up. The sun-baked red soil of the karroo is, however, the best travelling in the Colony, and differs widely from the agonies of mountain trekking, which are not only indescribable, but not to be imagined by users of English roads. We had not proceeded more than an hour, when, 300 yards in our front, a herd of fifty or sixty springbok galloped across, and cleared the road as they went in most beautiful style, each antelope bounding over like an india-rubber ball. Hurriedly we pulled up, seized our rifles, jumped out, and blazed away at the retreating game. As the bullets whistled past them, the springbok, not to disgrace their name, leapt high in the air, ten feet or more, arching their backs and unfolding the long, snow-white hair that usually lies hidden upon their croups. A singular and beautiful sight indeed! Again we fired, and this time the tell-tale thud proclaimed a hit, and a hard one, for an antelope staggered, fell, and shortly was secured; a good shot, and a most lucky one, at nearly 400 yards. We carried the dead beauty to our cart, and again trekked with renewed spirits.

Onward we toiled over the parched and now heated plains, outspanning at intervals of two hours or so during the day, seeing no sign of human life or habitation till late in the afternoon, when we reached a store and farmhouse.

At this outspan we met a gentleman, with whom we were afterwards better acquainted, Mr. David Hume, son of one of the Algoa Bay settlers of 1820. His father, also a David Hume, was the pioneer of British enterprise in the territory now called Transvaal—then a *terra incognita*, held by the fierce Moselikatse and his Amabaka-Zulu hordes. The first expedition of this gentleman to that country, undertaken in 1833, may be found marked upon the map at the end of Cornwallis Harris's delightful book, " Wild Sports of Southern Africa."

The elder Hume was also one of the earliest, if not the first British trader and hunter among the more northerly of the Bechuana tribes, as may be noticed by his route marked upon Harris's map. From Mr. David Hume, himself a great interior hunter, and a mine of information concerning the far interior, we obtained much interesting intelligence. He had made several trips to the Zambesi. Upon the last one, which he undertook in company with two English officers, he had barely escaped with life from the malignant Zambesi fever, to which his two companions had unhappily both fallen victims.

The life of an elephant and big game hunter in those regions is, indeed, one of incessant hardship and danger. And yet, such is its absorbing fascination, that once bitten with the *Cacoethes venandi*, these hunters will return season after season to renew the perils and delights of the dim and dangerous wilderness.

At this place we parted with the gentleman

who had guided us from Graaff Reinet, and now left to our own devices once more, we inspanned and set forth. We had before us a drive of between two and three miles before reaching Swanepoels Poort, where we should halt for the night. We got on well enough—barring the increasing roughness of the road—till dusk, when we had still some miles to compass before finding shelter. We had now some little difficulty in keeping to the rude track, miscalled a road; once or twice, descending suddenly and without warning into deep spruits, we had terrific shakes up, the cart all but capsizing. However, we blundered along as best we could, until at length we neared a light on the right hand. We were now on the banks of the Plessis River, a Karroo stream, which here runs through the narrow mountain gorge known as Swanepoels Poort. The drift or ford here is one of the most precipitous and dangerous in all the Colony, and even in broad daylight has to be crossed with great care. Of all this, however, we were in blissful ignorance, and finding ourselves on the brink of the river, we straightway prepared to cross it, and attain the light we could see burning just over on the thither side. As luck would have it, we took the drift in the right place—otherwise we should have tumbled cart and all over the high rock bank into the river—and very speedily found the horses scrambling and slipping down what seemed like the roof of a house. Here we were in deep water, through which we plunged, and then with a rush and shouts of encouragement, our gallant nags scrambled up the other steep and lofty bank. It was a most unpleasant sensation on a dark night, and if we had known what was before us, I don't think we should have tempted Providence. In fact this is one of the very nastiest and most dangerous drifts met with in the whole of our journeys, and as Mynheer Stols, the Boer to whose house we went, remarked to a friend of ours a day or two after, "It could only have been English *umlūngos** or drunken men who would have taken the drift on such a night."

We at once made our way to the house, and to our disgust found, after much trouble (for we could then speak no Dutch, and the Boer who came to the door very little English), that we had in the darkness missed the Englishman's house, which stood on the other bank of the river. Here was a pretty mess. We had now to make that horrible crossing again, and we didn't half relish the idea. However, there was nothing for it; if we wanted supper and a shake-down in comfort, we must return. This time I got out and went to the near horse's head, and with a good deal of snorting from the nags, who relished the business as little as ourselves, a long sliding flounder, a rush, and a scramble, we managed more by good luck than judgment to regain the other bank. For myself, I emerged from the operation wet to the middle. However, we soon found Mr. Boyce's house, and having outspanned, and seen the nags stabled and fed, sat down to a comfortable supper, and after a chat turned in. Our host, an old 72nd man, was Field Cornet † of the district, and gave us a good deal of information concerning the neighbourhood.

---

* *Umlūngo* is a very expressive Kaffir word, signifying a sort of gentlemanly know-nothing or green-horn.

† Field Cornet, a kind of deputy magistrate, who frequently acts as postmaster, and undertakes minor administrative offices.

Next morning, after discharging our reckoning, we prepared to complete the six miles that remained before reaching our destination in Naroekas Poort. We crossed the drift, which in the morning light looked even more ugly than we had thought it in the darkness of the preceding night, and then went up the pass that lay between two precipitous and forbidding walls of mountain. This grim Swanepoels Poort was, as we afterwards learnt, one of the last strongholds of the untamed Bushmen south of the Karroo. Up in the rocks that tower above it are their caves and fastnesses, and there may yet be seen the remains of their pictorial art—the strange presentments of the various beasts of the chase. Swanepoels Poort, which gives access from the Eastern Karroo, and the Eastern frontier of the Colony, to the Western province, was formerly on the main route of travel. The Bushmen, perched secure in these mountain eyries, and armed with poisoned arrows, were a constant source of danger and annoyance to travellers passing through; and, moreover, they harried the flocks of the neighbouring farmers incessantly. But one day, about the year 1830, Sir Andries Stockenstrom, Lieutenant-Governor of the Eastern province, one of the finest specimens of the colonial warrior and statesman, and rightly created a baronet for his manful and most eminent services in the old frontier wars, as well as in time of peace, passed through the poort with a considerable retinue. In his turn, annoyed by the Bushmen marauders still lingering there, he heard from a Boer living just outside the pass of their many misdeeds. "Well," said Sir Andries, "if you like to raise a commando and expel these robbers, I will procure you a large grant of land within the poort." The offer was accepted, a commando was raised, and the fierce Bushmen, fighting grimly to the last, were, after great trouble, all slain or expelled from the mountains that for centuries untold had given them shelter. Bushmen there are still in these parts, but they are at all events semi-civilised and quite harmless, and they now prove themselves the best of native servants, as hunters, herds, and grooms.

With our coaching-horn we woke the slumbering echoes of these grim defile as we passed blithely if laboriously through, and then for the remaining five or six miles we drove through some of the wildest and most sublime mountain scenery in the Colony. The gaunt and desolate masses of Witteberg heaved everywhere around in solemn and rather awful grandeur. Down in the valley, far beneath us, ran the river, fringed with bush and thorn, and here and there passing through groves of acacia. It was a magnificent drive, marred only by the frightfully rough travelling. Road there was none; long staircases of rock, down which we banged and rattled, huge boulders sticking up here and there in the middle of the track, and occasional deep holes, made things more than lively for us; and through all this had come our host and hostess and family, to take possession of their mountain farm. Every stick of furniture, every plank of wood about their house, all that made up their home, had come over these six awful miles of mountain track. Colonists at the Cape, especially they who pitch their tents among the mountains, have some rough experiences indeed. But at last, after one slight capsize, we reached the river-bed, and for the rest of the way up to Naroekas Poort (another mountain pass), where lay the house of our friends, we fared well enough. Finally, twanging cheerily upon the horn to let them know we were coming, we drove up the last steep bit of hillside, and pulled up in front of a good-sized, square, flat-topped, single-storied house, where for some time to come we were to sojourn. Our coming was unexpected, and our welcome more than a warm one. We spent a long enjoyable day, receiving and imparting news, and sat smoking and talking far into the night.

Chapter III.

KLIPSPRINGER SHOOTING.

I HAVE before mentioned that bedsteads, at all events those of European construction, are not so plentiful in the Colony as could be wished. The cost of transport by ox-waggon from the coast operates as an effective bar to the free introduction of articles of luxury, and even of necessaries, where the farmstead is remote, as is usually the case. In the last few years the Cape railways have done something towards remedying this state of things, but even now the area of the Colony is so vast that most localities, far from the trunk lines, still have their household supplies brought to them by the slow and expensive ox-waggon. H's farm in Naroekas lay seventy miles from the nearest village, or small town (Willowmore), and upon our arrival there we found but three iron bedsteads in the house. Of these, H. and his wife made use of one; the English nurse and the little girl filled the second; while Charlie H., our host's brother, had possession of the third, which stood in a large airy bedroom. In this room another bed was therefore made up on the floor for two of us, while the happy third member of our party—Frank (who, at "odd man out," had won the toss for choice of positions)—luxuriated in a half-share in Charlie H's double bedstead; whence, on opening his eyes next morning, he gazed contemptuously on Bob and myself, grovelling in our lowly " doss."

We were awakened at six o'clock by Sallie, the little Kaffir housemaid, who, entering noiselessly with bare feet, brought us each the usual early morning cup of coffee, without which no Cape colonist begins the day. A cheery breakfast party was ours, three quarters of an hour afterwards, and then, the inevitable pipes being lit, we discussed the programme, and it was settled that we should have a quiet look round that day, and hunt klipspringer on the following. Strolling out of doors, we three visitors once more gazed with curious and even wondering eyes on the strange scenes that lay stretched before us. Deeply set in the heart of the sombre yet magnificent Witteberg range, our home and abiding place for some months to come was a good example of a remote pastoral farm in Cape Colony. The name Naroekas, or Naroogas, is of Bushman origin.

Here, in a wild and sequestered valley, lived our friends, who occupied a mountain farm of 18,000 morgen—about 36,000 acres. The farmhouse was single-storied, roomy, square, and flat-roofed, and was built of Kaffir bricks and whitewashed. It lay on the mountain-side near a constant stream or fountain, and overlooked the pass called Naroekas Poort. The scenery around was of a wild and savage beauty; huge brown mountains, broken here and there into thickly bushed kloof and ravine, reared their heads on every side. Although quantities of lovely flowers and shrubs blossomed around us, the poort, from its sombre desolation and lack of human life, possessed a peculiar solemnity and gloom, and even on a hunting expedition this feeling could not be thrown off, until, after traversing many a mile, one gazed at length upon the open and far-extending Karroo plains. These rude hills had for centuries sheltered the little Bushmen hunters, the aboriginals of this strange land. In the old days, the Bushmen had no doubt wandered far and wide over the broad Karroo deserts, lords of the soil, and of the mighty beasts of the chase, just as they yet wander in the Kalahari and other game-frequented plains of the African interior. In time, however, the Hottentots had expelled the Bushmen, and driven them to the mountains, only to be themselves driven out by the Kaffirs, who, in turn, after many a fierce struggle, yielded to the Boers and to the British.

Grim and desolate as were these mountains, savage though the prospect around us, never while memory lingers can I efface the recollection of the delightful days spent on that remote mountain farm. Days of an existence, free, vigorous, and untrammelled; of hunter's toil, arduous, yet full of enjoyment, and healthful to an extraordinary degree; days replete with never-ending charm and change, spent amid a fauna and avi-fauna almost undisturbed, a nature almost as rude and untamed as it had been far back in the long centuries. It was an experience of pleasure altogether unalloyed; it is, and will be, a memory for ever fragrant. The great kindnesses received at the hands of our host and hostess, went far to heighten the charm of a lengthened and most memorable sojourn. We spent our first day in Naroekas Poort, in inspecting the horse and mule-breeding establishment in a distant kloof, as well as the flocks of Angora and other goats, and a few ostriches and odds and ends that made up the farm stock. It was arranged that evening that we should devote the next day to the klipspringers which abounded in these mountains.

Although all the larger and most of the smaller of the South African antelopes have been repeatedly noticed and described by numerous travellers, hunters, and naturalists—from Sparrman, the Swedish naturalist, who made an expedition to the Cape in 1772, downwards — the klipspringer, from its diminutive size and the inaccessibility of its habitat, has generally escaped notice, or has called forth but few passing remarks. The chase of the portly eland, the swift and enduring gemsbok, both lovers of the parched and open plains; of the noble koodoo, and the rare sable antelope of the mountains; and of all the goodly company of antelopes that grace South African mountain, bush-veldt, and karroo, has times out of number been extolled by enthusiastic hunters; but the klipspringer is, as a rule, little known to Europeans. Even at the Natural History Museum, and in the South African game trophy at the late Colonial Exhibition, the specimens are, and were, but poor. Le Vaillant, the French naturalist, who travelled at the Cape in 1784, certainly makes some slight mention of this antelope in his interesting journey to the Namaqua country, and gives its Hottentot name of "Kainsi"; but Cornwallis Harris, Gordon Cumming, Baldwin, Livingstone, Selous, Drummond, and, in fact, nearly every writer of note upon South Africa and its *feræ naturæ*, appear to have seldom or ever shot this beautiful and singular little creature; and if it be mentioned, it is usually

HEAD OF A KLIPSPRINGER RAM.

either to notice that its hair is good for saddle stuffing, or merely to include it in the ample catalogue of the antelopes to be found between the Zambesi and the Cape.

The klipspringer (*Oreotragus saltatrix*), or rock-jumper of the Dutch Boer, sometimes also called the klipbok (rock-buck), may be truly called the chamois of South Africa, and in many points closely resembles the chamois of Europe. It has the same marvellous facility for the retention of its foothold when leaping from one piece of rock to another, and in its powers of getting about the most precipitous and dangerous cliffs and mountain-sides, it certainly has no reason to fear comparison with its European congener. It will, when pursued, boldly leap from the giddiest heights on to the tiniest projection of rock, and its four hoofs will rest easily on a piece of rock not bigger than a large walnut. It differs certainly from the chamois of Europe in its horns, which in the male are about four inches in length, straight, and sharply pointed; the female being hornless. Its coat, too, is in its way unique, composed as it is of a thick growth of hollow, brittle hairs, and in colour giving as a whole the appearance of a rich olive brown which fades to grey underneath. The little klipspringer is sturdily built, and in size is about on a par with a three-quarter grown lamb. It stands some twenty-two inches high at the shoulder, and carries a beautiful full head, well set upon its sturdy neck, and its eyes are well placed, soft, and melting; its hoofs are somewhat hollow and curiously jagged at the edges. It is to be found on every mountain range in South Africa, but only in the most inaccessible spots, and it is exceedingly shy. The slothful Boers seldom, if ever, hunt it; consequently its only foes are British settlers, eagles, leopards (which abound wherever the klipspringer is found), and an occasional Kaffir or Hottentot.

When I first set foot in South Africa, I had scarcely even heard of this charming antelope, but in Naroekas I soon came to know and admire the little beauty. On the following morning, after a plentiful breakfast of goat chops and fry, honey and coffee, we started away in quest of klipspringers. There were four of us, and with us went Igneese, a Kaffir from a kraal or location down the valley, the best tracker and sportsman in the district, and a good fellow to boot. Picking our way down the rough stony hillside, we pass the little mealie and oat "lands" in the alluvial soil near the river-bed. Here we disturb, amongst other birds, a small flock of wax-billed grosbeaks (*Estrelda astrild*). This pretty little bird, common though it is, deserves passing mention. The Boers know it as the roodebec (red-beak), and, especially the corn-farmers of the Western province, are familiar enough with it, visited as they are by enormous flocks. Upon the back and top it is of a darkish brown colour, underneath a paler brown, in both places variegated by dark wavy stripes, and all over these prevails a most singular roseate purplish bloom. The bill is of an intense scarlet, which colour is also very notable in the eye stripe, and down the chest and stomach.[*]

---

[*] I have never seen their nests, but have often heard (and Mr. E. L. Layard, in his interesting work on South African birds, confirms my informants) that these birds nest in common, several using the same nest. When the united families arrive, there must be surely sorrow and headaches among the parents

The ruddy-billed intruders we had thus disturbed were, I suppose, searching for grains of comfort in the tiny enclosures devoted to the scant crops of our mountain farm. Now we crossed the bottom of the valley, and, after a two-mile walk, reached the foot of a mountain where we expected to find game. Igneese had so guided us that the gentle breeze blew in our faces, and we at once began the ascent. This proved a very serious undertaking, for the mountain-side was here composed of crumbling rocks and shingle, which were not only infinitely distressing, but necessitated the greatest caution as we approached a broad, bushy ledge, where our Kaffir first expected to hit upon the antelopes. Three-quarters of an hour's toil, however, found us well up the mountain-side, on fairly level ground, and not far from the ledge I have mentioned. We now separated and approached the cover in different directions, forming a third of a circle. In five minutes more we were in the bush, and very shortly, peering through it with the Kaffir, I espied two pairs of handsome little brown antelopes gallop rapidly from an open space, where they had been feeding, and dive into some bushes on the other side. We had no time to fire, but immediately rushed after the klipspringers. Two of them, which we followed, turned to the left straight up the mountain, while the other pair went

judging by the noise, shrill, and twittering they themselves give forth. Concerning the nest itself, Mr. Layard observes, " the nest is a large structure, composed of straw, grasses, feathers, wool, paper, rags, etc. It is often as large as a stable bucket round, and with an entrance in the side. The interior is a mass of feathers, and the eggs, from eight to fourteen in number, are pure white and oval . . . ." Imagine a nest common to half-a-dozen birds, each capable of laying fourteen eggs, especially if the hatchings out are coincident, as I believe they often are. The result must be surely appalling even to these light-hearted creatures.

right-handed, and were followed by the rest of the party. After several minutes' desperate scrambling up hill, through euphorbias, aloes, and thorny scrub, and over boulders, Igneese and I stopped for a blow, and took a cautious glance upwards. Instantly I beheld, one hundred yards above, perched on a rock which jutted abruptly from the mountain, one of the klipspringers—the ram—looking eagerly round for his enemies. I lost not a moment, but, with hasty aim, fired; and as luck would have it, I dropped the antelope with a bullet through the middle of the body, rather too far back from the shoulder. The klipspringer toppled headlong from his rock, and in breathless haste the Kaffir and I raced up to secure him. Before we could get up, however, the quarry was on his legs again, scrambling to denser coverts for sanctuary, though in vain. The Kaffir cries in Dutch "Schiet! schiet!" ("Shoot! shoot!"), and with the cry another bullet at thirty paces brings the bok to earth again, and the Kaffir's knife soon ends his struggles.

Meanwhile four other shots had echoed to our right, and fifteen minutes afterwards the rest of the party appeared with another klipspringer. We had been unusually fortunate thus early in the day, for, as a rule, these antelopes take an infinity of trouble to find, and still more to secure. Then we pursued our way, and reached the mountain top, which was flat, and covered with long waving grass. In a couple of hours more we sighted another brace of klipspringers; but this time, before we were within range, they were springing up the mountain cliffs like balls of india-rubber, and were soon far out of sight. Descending to a hollow in the hills, Igneese led us to a tiny waterhole, called Wilde Paard's Fontein (zebra's fountain), where formerly large numbers of zebras drank. These beautiful creatures were, at the time I write of, alas! reduced to a small herd of eight or ten, which still lingered in the remotest parts of the mountains in this locality. Greatly to our disappointment, for the sun was hot, and we were all athirst, this drinking place was dry; there had been a long drought, and the fountain now held nothing but hard-baked mud. Retracing our steps, we sought another part of the hills, and after a long scramble and another tramp along the flat grassy top of the mountain, we put up three more klipspringers that were feeding among some huge boulders littering the ground towards the edge of the mountain wall. The antelopes were off and away like lightning, and although Bob and Frank, who were nearest, each emptied two barrels at a hundred and a hundred and thirty yards, the odds were too great; the buck dodged in and out of the masses of rock in the most marvellous fashion, and were quickly over the cliff and far away.

It was a pretty sight, and we in the rear thoroughly enjoyed it, much as we sympathised with the rather exasperated gunners. Here we rested and ate a morsel, dry though we were, and as luck would have it, while looking down the hillside below us, we espied some water standing in the deep smooth fleshy leaves of a great aloe beneath. After some trouble, we managed to convey the contents of some of these leaves into the outside of the deep crown of one of our broad-brimmed felt hats, and thus partially slaked our thirst. Seated up here on the edge of the mountain, we commanded a magnificent view of kloof and hillside. Away near the cliff tops, an eagle or two could be seen sweeping hither and thither with ever-searching eyes. Close to us, on the right hand, hovered a stein valk (stone falcon), (*Tinnunculus rupicolus*), well known to Le Vaillant as *Le montagnard*. This pretty little kestrel greatly resembles our English kestrel, indeed its habits are almost perfectly identical. The reddish general colouring, the hovering flight, the swift-dropping stoop, and the quarry it follows, —small birds, mice, etc., are reproduced in "the mountaineer," precisely as they appear in its European cousin. The Boers call it as often rooi valk (red falcon) as stein valk. Presently we rose and resumed our tramp, with the result of bagging one more klipspringer, and sighting in the distance a herd of vaal (grey) rhebok (*Pelea capreola*), another mountain antelope. As we retrace our steps homewards, we pass through a kloof, where the sinking sun lights up the rocks on our left most gloriously, bathing them in the soft red, mellow glow of African evening, while on our right everything is in deepest shadow. Then, as we near home, we pause to watch the beautiful Angora goats drink at the pool below the house, the parting gleams of daylight lingering lovingly on their snow-white silky coats. Finally, after a twenty-mile tramp, just as a Kaffir brings in the game on a pony, with which he had followed us, we retire indoors to supper, afterwards to amuse our friends, as we had the Boers and accommodation-house keepers on our travels, with festive discourse on the banjo and a song or two.

Chapter IV.

LIFE ON A MOUNTAIN FARM.*

LIFE with the pastoral farmer in Cape Colony goes very quietly, very regularly—some people might even say something monotonously. But, to the enterprising farmer in South Africa, there is always plenty to employ the time; fencing, gardening, tree-planting, dam-making, well-sinking, these and a hundred other things beyond the ordinary pastoral routine, are for ever clamouring to be attended to. And to the credit of the too scanty element of British folk in the Colony, and in great contrast to the more slothful, and, if you will, slow-moving Dutch farmer, these improvements are ever and increasingly going forward, transforming barren spots in rugged mountain country or howling desert into cosy habitations and smiling farmsteads. Shut up in our lonely farm of 18,000 morgen (rather more than 36,000 acres) amongst the mountains, life runs quietly with us, it must be admitted. We are goat-farmers, and feed our flocks upon the zuur veldt (sour pasture) with which our rugged hillsides are clothed. I say "we" for the sake of convenience, for I, personally, and at present my travelling friends, are visitors. Some of our goats are of the ordinary kind, others are the beautiful Angoras—shapely creatures, with long white silky coats, from which is clipped the well-known mohair. The Angoras, as a rule, however, thrive better on the

---

* This chapter was originally written in the present tense, and I have thought it as well to print it in its early form.

Karroo plains, where they are pastured in vast numbers. We also breed horses and mules, which thrive well in these regions.

At six o'clock each morning—earlier in the summer—after Sally, our little Kaffir maid-of-all-work, has brought to each of us, while in bed, the matutinal cup of coffee, one of our number rises and goes to the adjacent kraal to count the goats as they issue forth in charge of their Kaffir herds to pasture on the mountain veldt. N'quami, our groom, a long lithe Kaffir from Galekaland, more commonly known as John, has slaughtered at the back of the house the goat which is each day sacrified for the needs of our household; and the kidneys, fry, and a chop or two are grilled for breakfast. Honey, found in abundance in the rocks around us, takes the place of marmalade. Breakfast over, one or two of us ride off to the broad shallow kloof, three miles farther away in the mountains, where our horses and mules are run, under the care of Tobias, our Dutch foreman. Another of us busies himself marking goats; whilst a third takes on his back a dead kid—well primed with strychnine pills—meaning to deposit it in a small kloof across the valley for the benefit of one of the numerous leopards which infest our farm. This is the only method of keeping down these night marauders, which do cruel execution on our flocks and foals. Then attention must be given to the little patch of alluvial land by the dry river-bed, where we grow sufficient grain for our wants, and odd jobs have to be done. A gun or rifle is usually carried; for game, feathered and furred, is plentiful, and is ever a welcome change from the monotony of goat-flesh. Meanwhile, Mrs. H., our kind hostess, is busy, superintending the washing, cooking, and other matters of domestic economy, as well as the education of her little fair-haired girl, our pride and delight. Kaitje, our cook, a yellow lady of mixed race (Hottentot, Kaffir, and Bushman intermingled), attends to the baking of our " cookies," cakes of unleavened meal, which usurp the place of bread, and which are baked—and capitally baked too—in a hole in the ground, heated in the native fashion. If we happen to be near the homestead, we take a light lunch in the middle of the day, and at seven o'clock we all meet for dinner; after which we smoke and chat, or read, till bedtime—usually about ten o'clock. If we do not make money fast nowadays, we live in a rude plenty: meat in abundance is ours; dried fruits—quinces and peaches, excellent when stewed—bought from the Boers who pass us occasionally with waggon-loads of produce; oranges and grapes in season, from the irrigated farms of Boers out beyond the poort, whence we emerge from our mountain home; and porridge of pounded mealies and milk occasionally. Goats' milk we have in plenty, but no butter or cream; for cows and oxen do not flourish upon our zuur veldt. Game abounds with us. Our mountains support several kinds of antelope—the klipspringer, rhebok of two sorts, vaal and rooi (grey and red), duykerbok, and an occasional koodoo; the latter one of the noblest of the South African fauna, carrying magnificent spiral horns, and weighing from 400 lbs. to 500 lbs. Francolins and bustards and other feathered game are also plentiful. A day's shooting forms a pleasant break in the quiet round of our existence. The chase of the beautiful antelopes that yet remain to us, the varied splendour of our mountain scenery —rich in glorious flowers—and a perfect climate, are indeed our chief pleasures; though, sad to say, they hardly compensate to us (I speak from the farmer's point of view) the lack of more sordid pelf.

> " Our rocks are rough; but smiling there
>   Th' acacia waves her yellow hair;
> Lonely and sweet, nor loved the less
>   For flowering in the wilderness.
> Our sands are bare; but down their slope
>   The silvery-footed antelope
> As gracefully and gaily springs
> As in the marble courts of kings."

Our home in Naroekas Poort is, as I have before mentioned, a single-storied messuage, built of Kaffir-made bricks, flat-roofed (as are most colonial farmhouses), and whitewashed, and perched high on the mountain-side, near a perennial fountain that issues from the rocks above. The fountain in South African farming is the mainspring by which everything is set in motion, for without a permanent water supply all else—the best veldt, the finest pasturage—is of no avail. Usually, having had early morning coffee, we stroll down to the angle of the river, just where it bends suddenly and runs through the poort. Here there is a deep pool, where, even now in time of drought, we can always count on getting a good plunge and a short swim. Just now, the river being low, the water is brackish; when the rains fall and the floods descend, it will become fresh enough for a time, only, as it falls again, to resume its salt savour. Sometimes, when the drought on the karroo is very severe, the waters of this river become so salt that the very fish perish, and are found floating dead in large numbers. These fish are a kind of *Silurus;* they wear long wattles, like our English barbel, and are called by the colonists, barbers. While we bathe, the baboons living in the rocks around come as near as they dare, and bark fiercely and angrily at our intrusion. This they do morning after morning, showing no softening of manners—no incipient sign of friendship. If, however, we have a gun with us, they are very careful to keep at a respectful distance, and behind shelter of their rocks, whence they quah-quah more angrily than ever; they are curious beasts, and most mischievous. If the farmer lays unction to his soul, and congratulates himself upon a more than usually fine piece of mealies (maize), or oats, or pumpkins, the odds are that one fine day he will find that the baboons have paid him a visit, and destroyed and torn up half his crop. Moreover, they do this in sheer and wanton mischief, befouling and destroying far more than they ever eat. For these reasons the farmers are their sworn foes, and, occasionally, as the only means of destroying them, a party having, by the aid of a Kaffir or Hottentot, tracked them to their sleeping places (usually large caves), will, by the aid of bull's eye lanterns, shoot a large number at one time. I have only assisted at one of these night forays, and I must confess the business is not a pleasant one. The brutes, lying in the far corner huddled together for warmth, are bewildered and half-stupefied by the lantern rays flashing in their eyes, and at the first discharge usually a number are slain. Then there is a rush for the entrance, and the stronger and older baboons often succeed in running the gauntlet, occasionally knocking a human foe off his legs, but seldom or never using their powerful jaws and teeth. It is a fact that the baboons are far cleaner in their habits than most people would imagine, shifting their sleeping quarters from cave to cave as they become foul or too much infested with vermin. Of late years, in the mountains bordering on the karroo, the baboons have developed a new and most troublesome propensity. Some years back, some one baboon having, by chance, come across the dead body of a milch goat, discovered and extracted the milk-bag (they are passionately fond of milk), and, like Eve, " saw that it was good." His discovery must have been quickly imparted to his fellows, for the Karroo farmers began to find their milch goats ripped up and slaughtered by these brutes solely for the sweet and luscious milk. This horrible propensity has grown to alarming proportions on some farms, and hundreds of goats have been destroyed in a season in this way. The baboons, too, becoming accustomed to butchering, presently turned their attention to the flesh, and will now destroy kids, and, if they can manage it, goats, for their flesh alone.* In these localities, a war of revenge has been waged by the enraged farmers, not altogether successfully, for the cunning simian murderers are hard of utter extermination. The Hottentots are full of amusing tales about the baboon, which occupies a large space in their folklore. If you want a funny story, you may always

---

* There is a curious similarity in this modern development between these murderous baboons and the sheep-killing parrot of New Zealand. In either case the evolution has been co-incident with the increase of the flocks. The methods are different—the result is the same. The baboon attacks the goat's milk-bag, the parrot the sheep's kidney—through the back. Death is effected in either case.

get one from a Tottie, on the subject of these creatures. Returning from our bathe, we have breakfast, and then set about the work of the day. Our farmhouse has not been long built—about three years. It boasts four really good rooms, spacious and lofty, besides a good kitchen. Owing to the difficulties and expense of transport, our furniture is plain and serviceable, and something lacking from an ornate point of view. Our living-room boasts a strong table, a plain sideboard, a bureau, eight or ten strong Windsor arm-chairs, half-a-dozen pictures, and a large chest or two. The ceilings are of deal; every inch of the timber of these and the flooring has been imported from abroad, and brought hither by ox-waggon from Port Elizabeth. Guns and rifles, and the heads and horns of mountain antelopes, help to adorn the bare walls, which are unpapered and plainly distempered; some cocoa-nut matting completes the furnishing of the apartment. In one corner lies a great pile of oranges. These we procure from an occasional fruit Boer passing with his waggon through the poort, and during the day this corner, with its golden store, is often visited by all of us. In another corner of the room lies a neat pile of Boer tobacco, dark stuff, twisted up into great rolls, weighing many pounds, fastened together at the ends by pegs of bamboo. From these twists we cut our supplies as wanted. Tobacco is, happily for the Cape colonist, extraordinarily cheap. We pay from twopence to sixpence per pound for the commodity—ours averages about threepence—and occasionally from the better class of Oudtshoorn Boers, we procure a really decent smoking article. You need to get accustomed to Boer tobacco, but the taste once acquired, there is nothing to grumble at. Occasionally, from the Bay (Port Elizabeth), we get, as an especial luxury, a supply of American cake tobacco, which costs from two shillings to three shillings per pound for the very best quality. The window of this room, which commands a magnificent view of the pass straight in front, and of the tall mountains that environ us, opens—as, too, does the principal bedroom—on to the stoep, or terraced verandah, the lounging and smoking place of every African farmhouse. From the stoep, a flight of steps run down to the mountain-side, hereabouts, and elsewhere round the house, cleared a little of the stones and boulders that litter the slope.

In the sideboard, and a cupboard of our host's bedroom, are contained the few bottles of whiskey and "French" (every Afrikander knows real Cognac as "French" in contradistinction to Boer brandy), which may be required for medicinal purposes, or an occasional luxury. One more confession of this spiritual kind. In the one large cupboard, which is fixed into an angle of the wall of our dining-room, and which I have omitted to mention previously, stands a small cask of the best Cango, from which we draw our very moderate supply of evening grog. Cango, I may explain, is the best kind of colonial-made brandy; it is of a rich yellow colour, is produced in the Oudtshoorn district, and when matured, is really a very reasonable substitute for the more expensive foreign liquors. The ordinary Boer brandy a colourless, and, usually, frightfully raw spirit, is a very different stuff, and has a most peculiar and indescribable flavour of its own. It is procurable at the rate of from sixpence to one shilling per bottle, and from its cheapness and its alcoholic strength, is a baleful source of drunkenness and injury, especially to natives, throughout the Colony. We have no cellar, and thus our host is compelled to keep his supplies of spirit within the living-rooms, otherwise they would, I fear, not long escape the attentions of the native servants about the place. Coffee is our staple beverage, and morning, noon, and night, accompanies our meals.

Our two native servant girls sleep in the kitchen, curled up on the matting in front of the hearth. They do not change their clothes for the night's rest—nothing, I suppose, would induce them to. Their costumes consist of nothing more than a print gown—usually blue, spotted or chequered with white, and the red and white spotted cotton handkerchief with which they hide their woolly pates. When these garments are dirty, they are made to change them, this is the utmost they can be persuaded into. They wear no shoes or stockings, and herein I think they are wise enough; their soft pattering footsteps make one annoyance the less in their management. They are a great trouble to our hostess, these black servants. Kaitje, although she can cook very decently, is the emptiest-headed, silliest, most tittering creature that ever cooked a stew. If you but look at her she giggles, for no earthly reason that one can discover. Occasionally she sulks, and nothing can be done with her. Sally, a young good-looking Kaffir girl from the kraal a few miles away, is really not so bad; left to herself, and the eye of her good mistress, I believe she would improve vastly. But the sniggering Kaitje demoralises her a good deal. Occasionally, too, Sally goes on strike, that is, an unconquerably lazy fit comes over her; she absolutely refuses to work, and will lie underneath a bush at the back of the house for the whole day slumbering in the sun. I suppose these fits are but the natural outcome of centuries of a savage existence; but they are very trying to the mistress, especially when she knows that if she dismisses the recalcitrant "slavey," she cannot obtain a better substitute, and may probably fare far worse by the change. In South African establishments, I am afraid it must be owned that the old Boer methods of strong and frequent corporal punishment are the only ones of much avail. It is certain that the Dutch are far better served by their native servants than are the British. On Saturday Sally is allowed "her afternoon out," and proceeding to the bosom of her kraal, returns next morning not by any means improved.

John, the Kaffir groom, and the native herds, sleep in some outhouses in the rear. Jackson, an English mason and handy-man, at present engaged completing the building of some stables, has a comfortable shake-down in the building he is erecting. One other person completes the establishment, though he is, like Jackson, only a temporary hand; this is a Basuto, who is building stone kraals for the flocks—hitherto principally folded for the night in thorn enclosures cut from the *Acacia horrida*. This Basuto is the merriest, most good-tempered, light-hearted, steadily-working soul that ever owned a black skin. He is, like all the Basutos, and many of the Bechuanas, wonderfully skilled at his work; and his kraals, albeit they are built of the loose stones he finds around, are models of strength and symmetry. All day as he works he sings loudly to himself. I have promised him a pair of second-hand military trousers, and he is almost too effusive in his gratitude and affection; yet I cannot help liking the merry, good-tempered soul. We jabber together occasionally, and laugh mutually over our lingual difficulties. I gather from this Basuto that his ancestors, a generation back, found their building capabilities of great assistance in protecting themselves from the attacks of *Moselikatse* and his destroying hordes of Amabaka-Zulus, then ravaging the Transvaal and adjacent territories.

In this and most other portions of Cape Colony away from the southern timber belt, great timber trees are unknown. In the bottoms of the valleys, dense groves of mimosa (*Acacia horrida*) are found, and in the kloofs and smaller ravines the wild olive (*Olea verrucosa*), the kaffir plum (*Harpepphyllum caffrum*), and very occasionally the cedar boom (*Widdringtonia junipernoides*), and the willow by the rivers; but the English eye, scanning the average up-country farm at the Cape, searches in vain for the tall trees and spreading foliage of the Old Country. Tree planting is, however, coming more and more into vogue, and will effect great changes in this respect. Our mountains possess a large variety of bush and shrubbery; there is a wealth of flowering and fruiting shrubs. The goats find upon the mountain-sides quantities of spekboom and other fattening plants; as for flowers, there is no limit to them. Heaths, orchids, lilies, amaryllids, irids, wild geranium, and pelargoniums, in particular, are to be found in astonishing profusion. Aloes, euphorbia, and other cactus-like plants abound, and here and there that pest of the Cape farmer, the prickly pear, with its golden delicious-looking fruit, flourishes all too well. The prickly pear, that Mexican marauder, is not indigenous to South Africa, and no one seems to know quite when it first made its appearance. It propagates and spreads itself with alarming readiness, and is most difficult of extermination. The mellow-looking fruit is covered with a dense growth of the tiniest, most silky-looking spikes; and, once the confiding stranger allows these to come in contact with his skin, he is not soon likely to forget their poisonous and insidious properties. If these spikelets are scraped from the fruit, it can be eaten without further trouble, but it is a disappointing *morceau*, when all is said and done. Cattle and other stock that once take to eating this fruit, as they will do in time of drought, suffer frightfully; their mouths become literally masses of inflammation, and death is not seldom the penalty. Upon some farms, this baleful plant has spread to such an extent as to make the land almost worthless, and once it attains a strong hold, it is well-nigh impossible of extermination, neither fire, poison, nor the knife effecting its downfall. One of our great pleasures is to ride up the long deep kloofs that pierce the mountains, and inspect our horse and donkey-breeding establishment; this lies some miles away from the house, and the stock wander over a large extent of valley, ravine, and mountain. Much of the lower parts of these hills is clothed with rooi-grass (red grass), in which the rooi rheboks, following that curious protective instinct of nature, find shelter and feeding ground. Their cousins, the vaal (grey) rhebok, are to be sought amid the long grey grasses higher up the mountain. We have no lack of cattle to ride; if we want a fresh nag we have only to have a few driven down and take our pick, and after a week's grooming, the bush-ticks are cleared from their skins, and they begin to present a decent appearance. Sometimes we visit the Kaffir location, or the Boers farming outside our mountain passes— Swanepoels Poort in one direction, and Witte Poort, our alternate exit, in the other.

" 'Twas merry in the glowing morn, among the gleaming grass,
  To wander as we've wandered many a mile,
  And blow the cool tobacco cloud, and watch the white wreaths pass,
  Sitting loosely in the saddle all the while."

Following the Plessis River through our poort, and riding some four miles alongside of its alluvial bottom, the Kaffir kraal I have spoken of is arrived at. Here, located upon some Crown lands, is a considerable encampment of Amakosa Kaffirs, once the bitter foemen of the colonists, now harmless enough. These people, some sixty or seventy in number, live much as they have always lived, in a semi-tribal state. They have their petty chief, who is a friend of our host and a very decent fellow, and they live quietly enough, pursuing a lazy pastoral existence, in which the women-folk do most of the hard work. It is a pity they won't work, but they have the greatest repugnance to enter service with the colonist, save now and then intermittently—a few of them—as herds. Their huts are enclosed within a ring fence of thorns, and their kraals are near at hand. The chief's wife, N'Sana, a really handsome well-set up woman, is often to be met tending her husband's goats, her baby fastened at her back, a blanket carelessly yet most gracefully folded round her and over one shoulder. Occasionally they have some assegai practice, and it is interesting to watch them. Their powers of throwing this weapon have been greatly exaggerated to my thinking: at more than thirty yards it is odds on the moving target, and beyond seventy even a stationary object is pretty safe. The shaking, quivering motion they impart to their spears before casting them, is very curious. With the knobkerrie (knobstick) they are wonderful shots, knocking over with ease a running hare or a flying game-bird. The hunting grounds of these people, once lords over much of this country, are now greatly circumscribed; they can only pursue game upon their own lands, and it must be a sore point I imagine for them, when they remember that over these broad mountains, and the spreading plains beyond, their forefathers once hunted the gallant game wheresoever they listed. With Pringle's Captive of Camalu, they might sing:

> "O Camalu, green Camalu!
> 'Twas there I fed my father's flock,
> Beside the mount where cedars threw
> At dawn their shadows from the rock;
> There tended I my father's flock
> Along the grassy margined rills,
> Or chased the bounding bontebok
> With hound and spear among the hills."

On the whole, however, these Kaffirs are contented and happy, and some day in the future, they or their descendants will doubtless become more useful members of the social system of Cape Colony.

In the soft alluvial soil near the river, one often passes the huge holes of the aard-vark, the Cape ant-eater. Indeed, more than one part of the road itself has been occupied by the cavernous excavation of one of these singular creatures. The aard-vark never appears, except by the merest chance, in the day time, but at night, when the moon is up, you may sometimes, if you gang warily, happen upon him as he demolishes the tall ant-hills, and with long slimy tongue obtains his succulent supply of food. It has often puzzled me to understand how the aard-vark can live, as he undoubtedly does, for hours together—if pursued—completely shut off from the outer atmosphere. How does he breathe, —how sustain life? I once with some of our party essayed to dig out an ant-eater, but after labouring for an hour and a quarter on a hot morning, we retired vanquished. The beast, with his wonderful claws and legs, digs far faster than the human delver. He will actually, if surprised, put himself underground, while he is being approached, unless, as very rarely happens, he is completely taken unawares. Our Kaffir hunter, Igneese, brought us in a dead aard-vark, which he had circumvented and killed with a blow or two on the snout from his knobkerrie. Cavernous as are the entrances to their holes, they rapidly taper down and twist off at right angles, and it would certainly puzzle the most assiduous digger that ever ferreted an English hedgerow, to reach the Cape ant-eater in its earthy dwelling, even though he laboured from morn till eve of a long summer's day. I only knew of one instance of a successful digging out. This feat was accomplished many years ago by my late friend, Mr. J. B. Evans, on his first arrival in the Colony. In all his undertakings, this gentleman was the most determined colonist I ever met with, and very naturally his determination invariably gave him success. In this instance he and some native servants started digging at an aard-vark about nine o'clock one morning. The animal, when followed in this way, as it digs, blocks up the exit, and with its strong feet pads and packs the earth firmly behind it into a dense wall. When this process goes on, as it did in this case for many hours, and the animal is completely earth-bound, it is extraordinary how it can exist apparently without a suspicion of air. My friend and his natives dug and dug, and the ant-eater pursued his tactics, turning hither and thither yard after yard, hour after hour, until in the end the persistence of the attackers had its reward; and about five o'clock in the afternoon the aard-vark was dug out quite dead. I suppose the lack of air and the fearful exertions it had undergone had at last killed it. The distance burrowed and dug was, if I remember right, on this occasion, something like thirty-five yards, measured along the zig-zags. The strength of this animal—resistive, not aggressive—is enormous. Two or three men may get a hold of its tail and hind-quarters as it enters its hole, and yet it shall escape them. The Boers of our district have a legend, that a stout ox-reim (hide-rope) was once fastened to the tail of a burrowing aard-vark, and then tied to the tail of a horse, with the idea of easily extracting the "earth-pig" like a periwinkle from its shell. Sad to relate, the tail of the ant-eater proved more sinewy and resisting than the tail of the horse, and the latter came out of the contest minus his caudal appendage. I will not vouch for the truth of this story, for the Boers, like other people, sometimes allow their imaginations to get the better of them.

Wandering along near the quiet river, often but a chain of small pools, many rare and interesting birds may be noted. The charming little plover (*Charadrius tricollaris*) may be often seen, with its greenish-brown coat, white forehead, and black and white collar marks (two of the former are of the latter colour), and white stomach.

The Kaffir crane and the Stanley or blue crane, the common heron, the beautiful purple heron, the hammerkop (*Scopus umbretta*), the bittern (*Botaurus stellaris*), called by the Boers "roerdomp," are seen pretty frequently. The blue rail (*Rallus cærulescens*), with its ruddy brown colouring, bluish-drab breast, and black and white striped sides, a bird nowhere very common in the Colony, was once or twice shot during my stay.

The red-billed teal, smee eendtje of the Boers, and the geelbec (yellow-billed) duck (*Anas flavirostris*), are not uncommon, while very occasionally we come across the big berg gans (mountain goose) (*Chenalopex ægyptiacus*), a magnificent fellow, whose harsh noisy "honk" warns us of his whereabouts. When secured with a charge of swan or buckshot, this wild goose is a notable prize. The painting of his plumage is very fine. Greyish-red upon his top and under parts, strongly marked between the pinions with black, brick-red as to his wings, which are further variegated with white and bright green markings, black rumped, and further painted with splashes of deep-red upon his breast and round the eyes, and having a ruddy circle round the neck, one cannot sufficiently admire the form and splendid colouring of this noble bird.

Thus, with these and other wild fowl, too numerous to catalogue at length, and the various francolins and bustards, of which more hereafter, we have good store of "feather," as well as "fur," with which to satisfy those cravings for sport implanted in the breast of every one of us.

But stroll where one will about these quiet valleys and over these rugged hills, there is ever something to interest one. Of small birds; of the thrushes and starlings, larks, finches of innumerable species, butcher-birds, shrikes, fly-catchers, warblers, swallows, buntings, cuckoos, wheat-ears, colies, lories, honey-guides—which are included in the great family of cuckoos—and other varieties, we are for ever encountering some new specimen; and it would occupy all the time of a professed naturalist, in a year or two of ardent collecting, to enumerate and classify the many beautiful feathered creatures we see around us.

A notable starling, the green spreo (*Juida phœnicoptera*), is often to be seen in large flocks; its colouring is very beautiful. The head, shoulders, tail, rump, and upper parts of the legs, are of a vivid bluish-green; there is bright violet colouring over the ears, and the shoulder coverts are decorated with a showy stripe of red and violet. Once or twice in our rides and walks we noticed the curious orange-shouldered bunting, *Vidua phœnicoptera* of Swainson, sometimes also called the Kaffrarian grosbeak. The male of this species is famous for its immensely long tail; his colour is black, lit up by deep-red shoulder markings. The female and,

I believe, the young male bird is yellowish-brown in colour, has black wing-feathers, and is rendered conspicuous by the brilliant orange shoulder-patches from which the bird takes its name. Mr. Layard, in his "Birds of South Africa," mentions the difficulties the male bird of this species labours under when he has donned his breeding plumage, by reason of the inordinate length of his tail. Children are enabled to run him down; he cannot fly against the wind, and in wet weather, knowing his helplessness, he stirs not out of the bushes that hide him. Moreover, the Kaffir children stretch well-limed lines across the fields; thus numbers of these birds are snared by their tails becoming entangled in the sticky lines. I have frequently heard these statements verified by residents of the Eastern province and Kaffraria. Here is one more striking example of the troubles of matrimony, from a male point of view. The male orange-shouldered bunting, in his breeding plumage, no doubt, freely admits to himself that, in his case, "marriage is a failure."

There are one or two families of Boers residing like ourselves within the deep recesses of the Witteberg. They are of the poorer class of Boer; their holdings do not extend beyond the usual Dutch farm of a little over 6,000 acres (3,000 morgen), and, buried in some far-off recesses of the hills, their existence is even more isolated than our own. Sometimes we have to call in upon them in the way of business, sometimes we ride out to Swanepoels Poort and see the one or two families of Stols settled there. They are a strange, taciturn, rugged race, but when you know them, and if you approach them in a proper spirit, you may obtain a store of quaint information, grim legends, wild beliefs, and a fund of old-world hunting stories, handed down from their fathers, who at the beginning of this century hunted the elephant, rhinoceros, lion, and other great game that then abounded in the Eastern province. Too often the Boers are laughed at and ridiculed by the English settlers, and in consequence, they usually shut up like a knife when in their company. My friends here and at Riet Fontein pursued a different course, I believe a good deal to their own advantage in the long run. The Dutch are better as friends than enemies, and there is nothing to be gained by treating them as an inferior and even a conquered race—which latter they are not, nor ever were. Of late years, since they have discovered their political power, and, too, since the Transvaal War, they have become, even in the Old Colony, far more independent and even assertive than formerly. The habits of these strange people are wonderfully primitive. The occupants of a neighbouring farmhouse in the mountains are two brothers; one of them is married, but they have only one apartment, which is shared in common by night as well as by day, by the married couple and the single man. This is not an isolated instance, or one thought at all extraordinary among these simple folk. Father, mother, and grown-up children not seldom share the same sleeping room. In a subsequent chapter I have given some account of the Boer of to-day, who, as a matter of fact, resembles very much indeed the Boer of 1789 or even 1689. It is possible, and indeed I think highly probable, that the influx of the mining element into the Transvaal will induce far greater changes, and a speedier transformation in life and thought among the Boers there than in the Cape Colony, which for the present lies unnoticed by the invading British element, in the rush northwards to the gold-fields.

The Boers are superstitious to a degree. One of the Stols's, living out beyond Swanepoels Poort, was for some time driven almost out of his mind by what he considered ghostly manifestations. At all times of the day, and especially at meal times, stones, dung, and other filth were thrown into his house; windows were mysteriously broken, strange noises were heard, stones and garbage came flying on to the table at meals, and yet nothing could be discovered as to the origin of the manifestations. At last the poor man became so worried and nervous about the "spook," as he called it, that he went almost in tears to our neighbour at Riet Fontein—the late Mr. J. B. Evans—and implored his assistance as a magistrate. Mr. Evans sent to the nearest town for a policeman, whom he quartered at Stols's farmhouse, rightly thinking that the perpetrator of these high-jinks would be shortly discovered. For some time the policeman watched in vain, so well was the game played, but at length the culprit was unmasked. The eldest daughter of the farmer himself was caught *in flagrante delicto*, and confessed that she, and she alone, had wrought all this miserable deception upon her own parents. Her punishment at the hands of the enraged father was, I fancy, pretty exemplary, and the ghostly visitations never again returned. It is a curious thing that this peculiar species of deception, more common at one time among English servant girls than perhaps any other class, should have been thought of by this quiet Dutch girl upon a remote South African farm, who had almost certainly never heard or read of any precedent for such conduct. In a later chapter (xx.) will be found a Boer ghost story, sufficiently thrilling in its nature, and firmly believed in by the farmers of the district in which the scene is laid.

CHAPTER V.

THE ZEBRA IN CAPE COLONY.

THERE is apparently so much uncertainty as to the present distribution of the larger wild animals in Cape Colony, and so many people appear to be sceptical as to the existence of any game heavier than the springbok within its present limits, that I am induced to write a few lines upon a subject in which I take considerable interest, collected partly from personal experience, and partly from that of friends now residing in the Colony. To begin with, I may mention that of all the magnificent large game which have abounded within the Cape Colony within the last century, the elephant, the buffalo, the koodoo, the gemsbok, the hartebeest, the zebra, and the leopard are the sole representatives; and, with the exception of the leopard, which is still abundant, these animals are very seldom met with, and only prolong a precarious existence in the densest coverts, the remotest mountain tops, or, in the case of the gemsbok, the most waterless deserts of the western portion of the Orange River region. The rhinoceros, the hippopotamus, the giraffe, the quagga, the lion, the black wildebeest or gnu, the roan antelope, the eland, and other noble game, have all been exterminated or driven out of the Colony at a remote or near period of the last hundred years, in proportion to the measure of their several means of defence or escape. Thus the unwieldy rhinoceros and hippopotamus, and the defenceless giraffe, were first exterminated, while most of the remainder have only quitted the scene within the last generation. I say that the hippopotamus is exterminated; for, although some are still to be found in the western waters of the Orange River—the northern boundary of the Colony—in the rivers of the Colony proper they are undoubtedly extinct. I have mentioned in a former chapter on Klipspringer Shooting that the zebra (*Equus montanus*) still lingered in the wild and remote mountains surrounding Naroekas Poort. On the arrival of my friends and myself in the poort, I was surprised to learn that a troop of zebras still frequented this district; and, further, that these beautiful creatures were by no means of such infrequent occurrence in the Colony as we had imagined. My imagination was strangely kindled at the news; for I had believed that the true zebra had, with its cousins the quagga and Burchell's zebra, been long since exterminated or driven by the tide of colonisation into the interior. Ever since that time I have carefully followed the history of this particular troop, and, as far as possible, of its congeners in other parts of the Colony, and perhaps the result of my observations may be useful.

I believe that nineteen people out of twenty have a vague idea that the zebra is an animal confined exclusively to the wide karroos and rolling plains of Southern Africa.

The true zebra, the *Equus montanus*, *Equus* or *Asinus zebra*, the hippotigris of the ancients, the daow of the Hottentots, and the wilde paard (wild horse) of the Cape Dutch, is purely and essentially a mountain-abiding animal. It inhabits the most remote and rugged ranges of the Cape Colony; and at the present time, though sadly reduced in numbers and in the limits of its occurrence, it may be found in the Sneeuwberg, the Witteberg and Zwart Ruggens, the Zwartberg, and Winterhoek Mountains, and one or two other localities in the Eastern province. Quite recently a troop was running on the slopes of the Cockscomb, the highest peak (7,000 feet in height) of the Winterhoek.

The zebra, as all naturalists are aware, differs widely from the zebra of Burchell (*Equus Burchellii*), (the bonte quagga of the Boers, the piitzi of the Bechuanas), and from the quagga (*Equus quagga*) or quacha of the Hottentots. In the first place, its colour and markings are widely different. The zebra's body colour is of a beautiful silvery white, and the black markings are distributed more evenly, and extend to every part except the stomach and the inner parts of the thighs; even the legs are perfectly ribanded in black and white. Upon the light clean head (with the exception of the ears, upon which the black and white markings continue) the markings change to brown, while the muzzle is of a rich bay tan colour. The ears and tail are distinctly asinine, while in both Burchell's zebra and the quagga they are of the equine type. In height, too, the zebra, which only averages some twelve hands at the shoulder, is inferior by some six inches to its congeners. Contrasting, therefore, the colour and markings of the true zebra with the sienna body colour and brown striping of Burchell's zebra, and the reddish-brown of the quagga—whose stripings, by the way, only extend to a little behind the shoulders—and further noticing its distinctive asinine type when placed side by side with the more equine look of the other two, wide differences will be readily distinguished.*

The quagga and Burchell's zebra, too, are essentially lovers of the wide and open plains, while the zebra proper, as I have pointed out, never by any chance leaves its mountain home, unless possibly to trek from one neighbouring range to another, and then only under cover of night. I may remark here, that the quagga has only become extinct in the Cape Colony within the last twenty years, but Burchell's zebra must have disappeared long previously. That the latter animal did exist south of the Orange River in large numbers I have little doubt, from the perusal of Lieut. Paterson's very interesting "Journeys to the Country of the Namaquas and to Kaffraria in 1777-8-9," for he speaks of what was evidently Burchell's zebra being met with in large numbers in the Namaqua country, though at that time the animal was unconnected with the name of Dr. Burchell.

The first appearance of the zebra in history would seem to have been in the reign of the Roman Emperor Caracalla, who caused to be displayed in the circus an elephant, a rhinoceros, a tiger, and a hippotigris. That it was rare even to the Romans, those diligent collectors and wanton destroyers of animal life, is certain, for there seems to be no other mention of the "tiger horse" in the annals

---

* It is to be remarked that a rare variety of Burchell's zebra (*Equus Burchellii—var: Chapmanii*), completely striped, even to the legs, has been observed by Chapman and one or two other travellers in Southern Africa. This variety is, however, distinctly scarce, although I believe occurring rather more plentifully in Central Africa. A new variety of the true zebra (*Equus Grevyii*), from Shoa, North Africa, was also established in 1882.

of Imperial Rome. Further, it would seem that the zebra was unknown upon Egyptian monuments before Caracalla's time. Considering that the true zebra occurs in the mountains of Abyssinia, this fact is curious, to say the least of it, and goes to prove that *Equus montanus* has ever been a shy and difficult animal to obtain. The name zebra would seem to be of negro or Abyssinian origin, and is variously derived from Zeuru, Zeora, or Zecora— Abyssinian and Galla names.

Passing to more modern times, I find the first mention of the zebra made by Tachard, a Jesuit Priest, who touched at the Cape on his voyage to Siam soon after the middle of the seventeenth century. The early Dutch settlers called the zebra indifferently wilde paard or wilde esel (wild horse or wild ass), and the old world travellers fell into some absurd mistakes in consequence. Tachard gives a ludicrous wood-cut of these so-called wild asses, which from the full stripings are evidently intended for the true zebra. The stripings appear all over the body, but the ears are hideously misshapen, and more resemble cabbages or cauliflowers than anything else. Tachard may have seen a skin of the zebra, but his descriptions of these wild horses and asses are best represented in his own words. "There are both horses and asses here of extraordinary beauty. The first have a very little head and pretty long ears; they are all covered over with black and white streaks (here the true zebra is evidently intended) that reach from their back to their belly, about four or five fingers broad." From this fairly true sketch, probably taken from a skin,

Tachard passes into the realm of startling fiction. " As for the asses," he says " they are of all colours. They have a long blue list that reaches from head to tail; the body being like that of the horse full of broad streaks—blue, yellow, green, black and white, all very lively." Lively indeed! the word is all too poor for such magnificent colouring! Tachard calls his picture the zembra or wild ass, zembra evidently standing for zebra. The tail of the animal thus depicted is asinine and fairly typical of the true zebra. Ten Rhyne, who voyaged to the Cape in 1673, is nearer the truth when he speaks of this animal as being streaked with white all over.

Kolben speaks of the Cape wild ass as one of the most beautiful, well shaped, and lively creatures he had ever seen, and he tells us that it resembled the ass in nothing but its ears (here the true zebra is surely intended). " His legs are slender and well proportioned, his hair soft and sleek. There runs along the ridge of his back, from mane to tail, a black list, from which on each side proceed streaks of white, blue, and chestnut colour, meeting in circles under his belly. His head and ears, mane and tail, are also adorned with small streaks of the same colours. He is so swift that no horse can keep up with him, and as he is hard to be taken he bears a very great price." Kolben, although he sojourned at the Cape from 1703 to 1714, evidently knew the zebra only by report, and like Tachard when he condescends to particulars, flounders hopelessly out of his depth, in his effort to impart local *colouring*. His wood-cut gives the stripings all over the body, but the portrait is evidently an imaginary one; the figure (although striped) is the figure of a cobby horse, with a horse's mane and tail.

According to Tellez, the Great Mogul gave 2,000 ducats for a zebra. But Nanendorf relates that the Governor of Batavia having sent one of these animals (presented to him by an Abyssinian Ambassador) to the Emperor of China, that monarch sent the Dutch East India Company in return ten thousand " Taël " of silver and thirty night gowns, valued altogether at 160,000 crowns.

Such briefly is the lore I have been able to collect concerning this most interesting, and even now, almost unknown animal.

The first reliable portrait of the true zebra occurs in " Brooks' Natural History," published in 1760. In this curious old work (a copy of which I have by me), full of errors though it is in many places, there is a really striking one of *Equus montanus*. This wood-cut appears to have been taken from a living specimen at Kew, formerly in the possession of Frederick, Prince of Wales, father of George III. Although the picture apparently indicates the true zebra, I cannot reconcile with it the description given in the letterpress, which reads more like a description of Burchell's variety. The author has, I think, confused the varieties, and has taken Burchell's zebra to be the female, and the true zebra to be the male.

In the year 1887, a young but mature true zebra was captured in the mountains of Achter Sneeuwberg, in the district of Graaff Reinet, and was afterwards photographed. I obtained, with some difficulty, a copy of this photograph, which is here reproduced. So far as I know, this is the only representation of a mature true zebra captured in the wild state, and for this reason it is a curiosity. As a most faithful presentment of a fine specimen of an animal, always somewhat rare, and extremely difficult of capture, and now fast disappearing from Southern Africa, it is doubly interesting. I have to thank Mr. H. Roe, of Graaff Reinet, for his permission to reproduce this excellent likeness. Dr. Sclater, Secretary of the Zoological Society, was good enough to show me, a short time back, some photographs of zebras, from the collections of his own and various foreign societies. In one of the latter, I noticed with interest a portrait of a new variety of the true zebra, obtained from Shoa, North Africa, in 1882. This animal, mentioned in the proceedings of the Zoological Society for 1882, was sent alive by King Menelek of Shoa, as a present to President Grèvy, of the French Republic, and for a short time lived in the Jardin des Plantes. Although in some points, and especially in its markings, different from the true zebra of South Africa, it is, I think, undoubtedly a variety of Equus zebra. In this animal, the markings are finer and closer together than in the South African zebra, and in place of the broad horizontal bands running across the rump and thighs to the flank and ribs, so noticeable in the picture of the Sneeuwberg zebra, the stripes are much thinner and finer, and, indeed, vary but little from the other markings of the body.

A glance at the photograph of this new variety, gives one the impression that the stripes are all but absolutely even in width over the whole of the body. The nose of *Equus Grèvyii* is longer, more Roman in type, and altogether uglier than in the South African zebra, and the body and limbs are not so compact and robust.

There is, in fact, a more "leggy" look about the animal; the leg stripes are thinner and more numerous, and upon the top of the rump from the end of the back to the tail, on either side of the black list running down the centre, is a good sized space of white; this white space is absent in the true zebra. The ears of *Equus Grèvyii* appear to be shorter than in the South African animal. Beyond these points, there is little difference between the zebra of Shoa and of South Africa, widely sundered as are the habitats of the two animals.

The ferocity of the zebra is even more remarkable than it is in its relatives of the plains. The older Boers can give numerous anecdotes, illustrating the dangerous and savage nature of the animal, and, except when taken very young, it is perfectly untamable. In bygone days, when much more numerous than at present, I believe a considerable number of young zebras were captured by the Dutch, and exported to Mauritius, where they were trained to harness, and considered to impart an air of *ton* to fashionable equipages. Pringle—an acute observer and delightful prose writer, whose poems on South Africa, its fauna, and its natural beauties will live long after the last of the magnificent beasts he describes have been swept away—mentions the frightful injuries sustained by a young Dutch colonist from the bite of a zebra. The Boer was hunting in some mountains of the Graaff Reinet district, and had forced a zebra to the very brink of a precipice; the desperate brute turned to bay, attacked his pursuer with its teeth, and actually tore his foot from his leg, inflicting such a dreadful wound that the Boer died within a few days. Mr. (afterwards Sir John) Barrow, too, in his excellent book, "Travels into the Interior of Africa," published early in this century, writes of a soldier who had mounted a half-domesticated zebra in the Cape Colony. The creature, after making the most desperate attempts to get rid of its rider, at length plunged over a steep bank into a river, and finally threw the soldier heavily as it emerged. While the man lay half stunned upon the ground, the zebra quietly walked up to him and bit one of his ears clean off. The amusing Le Vaillant, in his "Journey to the Country of the Namaquas," about 1784, mentions having ridden a very fierce zebra, which no other person could manage; but in this particular instance his veracity is, I think, open to doubt. Interesting and instructive, and true to nature, as are very many portions of Le Vaillant's works, especially those pertaining to ornithology, some of the adventures are undoubtedly a little over-coloured by the vanity of the volatile Frenchman. But to return to the zebras of Naroekas Poort. I soon learned that the small troop of eight or ten, then running in the neighbourhood of our rugged pass, confined themselves to the most inaccessible slopes of the mountains, and that, except occasionally on long hunting expeditions after rhebok, klipspringers, and other shy game, they were never seen. I had the greatest curiosity to behold these beautiful creatures in their own wild fastnesses, and for many days, while following mountain antelopes, I looked far and wide for the richly striped wilde paard. At length one day, when out alone with

Igneese, the Kaffir, I caught a glimpse of the herd. I remember the day well. We had sallied out for a day's rhebok shooting, on a distant part of the farm, and after a long and unsuccessful tramp over some of the wildest mountains, and through some of the deepest and most lovely kloofs I ever saw in South Africa, we came to an abrupt corner ("hoek" the Boers call it) of a mountain, near to its summit. Stealing quietly round a sort of pass, the Kaffir suddenly whispered, or rather gasped, "Wilde paarden!" and I beheld, right in our front, and rather above us, standing on a rocky platform, a magnificent zebra, and a little beyond him six others. The troop was about two hundred and fifty yards distant, and for two or three minutes we stood motionless, regarding them. My host strictly preserved, as far as he could, these rare creatures; so, of course, shooting was out of the question, though the light in the Kaffir's eye plainly showed what his feelings upon the subject of preservation were. After a pause, we moved very stealthily forward, to get, if possible, a nearer view. In an instant, the sentinel we had first seen had discovered us, and at a wild, shrill neigh from him, the whole troop took to their heels, galloped headlong over the mountain top, and were quickly lost to view.

I had a fleeting glance at this troop on one other occasion, and I think, if time had permitted, I might have made closer acquaintance with them; but, shortly after, I left Naroekas Poort, in whose gloomy yet hospitable recesses I had passed so many happy days.

Since then I have heard occasionally from my friends in the poort and its neighbourhood, and a few months back I wrote particularly to inquire of the fate of the zebras, of whom mention had not been made for some little time. In reply to my inquiry, I was told that they had been reduced by Boers and natives, who never spare, or attempt to preserve, the lives of the rarer game of the Colony, until one stallion alone remained, the last representative of the striped beauties that from ages long remote had so fitly graced these rugged mountains. This stallion, finding himself the last of his fellows, had joined a troop of horses belong to the breeding establishment of my friends, that ranged far and free on the hillsides, and for some time ran with them. At length he became so accustomed to consort with the mares, that he followed them in all their wanderings, and finally suffered himself to be driven with them from the hills into a kraal or inclosure. It was then determined to keep him, and endeavour, if possible, to domesticate him. For this purpose he was lassoed and tied to a tree; but so ferocious and wild was he in the presence of man, that the greatest precautions had to be observed in approaching him, for his open mouth and ever-ready teeth were always prepared if any one ventured near. Every possible means was taken to induce the zebra to feed. When captured he was in splendid condition, and his coat shone in the sun "like a well-groomed horse's," as my friends expressed it. Herbage was brought from the mountain tops where he had been used to graze, and every conceivable food, equine and asinine, was placed before him, but in vain—he steadily refused to eat. Water he drank greedily, and would dispose of three bucketsful at a time. At length, after three weeks of vain endeavours to tame the magnificent creature, during which time he existed entirely on water, he died. It seems a pity that this noble brute should have been held captive so long, for he was a mature animal, seven or eight years old, and at that age it is impossible to tame the zebra.

Not long since I met an old friend, a former owner of land in the poort, who informed me that quite recently a fresh troop of zebras had trekked into these same mountains, where their spoor had several times been observed. That they may long remain with other zebras in the Colony to adorn these ancient haunts will, I am sure, be the wish of every sportsman and naturalist.

Chapter VI.

A RACE WITH A KAFFIR.

IT is a hot afternoon in Naroekas Poort; the sun plays steadily down upon the rough and broken sides, strewn with boulders of every shape and size, of the mountains that surround us; upon the white-washed walls of our flat-roofed farmhouse; upon thorn and shrub, spek-boom, prickly pear, cactus, aloe, and Kaffir plum that grow around; upon the little lands of oats and mealies down by the river-bed at our feet, and upon the kraals, some built of thorn, others of stone, lying to our flank, which at this time of day shelter only a few ostriches (for whom a more spacious camp is being prepared), who stalk hither and thither in foolish solemnity. It plays, too, upon the grove of thorny acacia (*Acacia horrida*) that clothes the bottom of our valley away in the hot distance, and upon the tiny clear stream —*fontein* we call it here—that trickles down the mountain at our back, watering in its course the small potato patch that has been won so hardly from these sterile rocks. Nature lies silent in the heat, the soft cooing of turtle doves and the song of the merry Basuto, building us a new stone kraal, alone disturb the repose. But why, the reader will ask, is your habitation pitched amid such rugged surroundings? My answer is—we are pastoral farmers, and these sterile looking mountains well serve our purposes. Our wide acreage of mountain and kloof of zuur veldt (sour pasture) affords excellent grazing for goats, and is admirably adapted for the breeding and rearing of horses. For this reason are we settled in one of the most sequestered nooks of the Eastern province.

Some of us have been marking goats in the kraal all the morning, while two others have been to the kloof across the valley, where a leopard has been crying for many nights past. With them they took a dead kid, into which had been deftly inserted, by means of incisions in soft parts of the flesh, a strychnine pill or two, intended for the benefit of *Felis pardus*. Arrived at the kloof, they had deposited the dead kid in a place where it was likely to attract the notice of the leopard, and in all probability the result would be the death of one more of these fierce enemies of our colts and flocks. Leopards are so essentially night-loving animals, that they are by the merest chance encountered in the daytime; hence the colonial method of attacking them by poison. Three years ago, when our host came to Naroekas Poort, it happened one day that the mason engaged in building the house was left alone for some hours; the walls had been erected, and he was working on the flat roof. Chancing to look down from his work, the man was astounded to see a leopard saunter quietly down the hillside and enter the house by the open doorway. Like most South African farmhouses, the dwelling consisted of but one storey, and the mason lay as still as death, for he was unarmed, and, knowing the activity of the brute, he was uncertain whether it might not scramble on to the roof if it caught sight of him. After five minutes of breathless suspense, the leopard, having finished its survey of the empty house, strolled out again and betook itself to the mountain, no doubt pondering inwardly as to the next piece of insolence that was coming to disturb the hitherto unbroken solitude of its fastnesses in the lonely poort. The mason was thankful when some more of his party arrived on the scene, and, needless to say, took care not to be left alone and unarmed at the house until it was finished. The leopard shares with the buffalo, at the Cape, the unenviable reputation of being the most dangerous animal to face when wounded. In this respect, even the lion, farther up country, takes a secondary place in the estimation of South African hunters.

Our hostess sits working on the stoep (verandah) just outside the dining-room, her little three-year old daughter is playing near her, and occasionally trots out to us with some treasure to display. She has no fellow-playmate; her child sister lies buried a little below the house, where a rude Kaffir brick grave marks the spot; for in such manner are Cape farmers, far from civilisation, often compelled to bury their lost ones.

As one gazes down the poort (or pass) in front of us, one cannot help feeling that these huge brown masses of rock that form the mountains of the Witteberg have a peculiar savage charm of their own. One's fancy wanders back to the time, not so many years ago, when the lion's roar echoed through their kloofs and kopjes, as the leopard's cry does even now; while the fleet hartebeest (*Alcelaphus caama*) and the lordly koodoo (*Strepsiceros kudu*), most beautiful of all South African antelopes, with its heavy spirated horns and rich striped skin, still graced the scene. Alas! that such noble and interesting animals should have been pushed back by the ruthless hand of man from these open valleys to safer haunts and more secluded pastures. Yonder in the valley is a pool in the nearly dry river-bed; it is called zee-koes gat (sea cows, or hippopotamus' deep), but the hippos have long since departed; whether from the attacks of human foes or from lack of sufficient water space, caused by that mysterious desiccating process so common to this country, is doubtful. The local names around us are mostly Dutch, though a few of Bushman and Kaffir origin still linger. I am not a thick-and-thin admirer of the Boers; but there is something in the strange and moving history of their sufferings, struggles, and wanderings in this land, in the quaint and uncouth, yet singularly graphic, names they have bequeathed to places and things, and in their simple Old World customs, that has for me a curious interest. These old names bring back to me scenes of long and weary trekkings, of outspans and inspans, of vast herds of game—now, alas! sadly reduced—of desperate fights with Kaffirs, and raids on Bushman hordes, too often, alas! sullied by deeds of unspeakable cruelty.

However, though one sighs for the bygone days, when this country was one vast game preserve—game the noblest and the worthiest of the hunter the world has ever seen—we must not grumble. There are still plenty of the smaller antelopes, such as rhebok (*Pelea capreola*), klipspringer (*Oreotragus saltatrix*), duyker (*Cephalopus mergens*), steinbok (*Nanotragus tragulus*), and on the Karroo plains, not far away, springbok (*Gazella euchôre*)—with which to amuse ourselves. As for francolin (here called partridges), guinea-fowl, koorhaan, and the rest, we can have as much shooting as we want amongst them. To-day, being somewhat of an off-day, and the warm sun of South African spring having a rather somnolent effect, we are sitting after lunch lazily smoking (except one of our number), and wondering what is to be done. Shall we stroll down to the pool at the bend of the river, half a mile beneath, for a bathe, or shall we stick up a bottle or two and have a little rifle practice?

The two proposals are ended by the distant rumbling of a waggon coming down the pass towards the house. Up jumps everybody, for be it known this is not an everyday occurrence, and in these quiet regions anything passing by the road below is an object of uncommon interest.

The waggon, as it rolls slowly along, anon crashing over a huge boulder, or across a dry spruit here and there, is soon made out to belong to Boers, probably from the Orange Free State, coming down country with produce.

The great oxen step out briskly as they catch sight of a habitation, and the Kaffir fore-louper at their head tugs at the *reim* by which he leads them. A flock of sheep and goats, headed and driven by Kaffir boys, accompanies the waggon, while a couple of useful saddle-horses follow it. We walk down the hillside to the road, and our host soon ascertains that the Boers, who have trekked far, desire to outspan for the night and rest the next day, and to run their trek-oxen, sheep, and goats upon the veldt; and after a little haggling this is accorded to them, in consideration of a sheep, which is forthwith picked from the flock, and taken up to our own kraal in charge of our Kaffir, John.

The Boer party consists of father, mother, two sons, and a daughter. The paterfamilias is a tall, loose-limbed, heavy man, of drabby-brown complexion, and sun-tanned, hay-coloured hair, and is clad in broad-brimmed felt-hat, decorated with a feather or two, short coat, flannel shirt, loose trousers, and veldt-schoons made of soft, home-tanned leather, with the hair outside. The sons are much of the same pattern, and they all bear a heavy somewhat sullen expression on their countenances.

The vrouw is a sunburnt, stout old lady, wearing on her head a tight cap or kerchief, surmounted by a large straw hat, and attired in a frock of heavy, rough material. Her daughter is rather more presentable, but is by no means of surpassing beauty. Her we do not see much of, as she sits far back in the waggon, shaded by the tilt. None of the party are of the cleanest; in fact it is a debatable question whether Boers, from the cradle to the grave, ever do indulge in the luxury of a bath— on the whole, probably, the " No's " would have it.

Their cargo consists of oranges, dried peaches and quinces, walnuts, Boer brandy, and Boer tobacco. They appear by no means eager to trade, and we have great difficulty in getting them to open the great hide bags in which they keep part of their stores. However, after sampling the brandy, things become a little brisker; we purchase some 200 oranges, a quantity of the tobacco at threepence per pound, and some brandy and dried fruits. Our host and his brother, as conversation becomes more lively, and a little banter is interchanged, draws the Boers considerably, by inquiring how their brethren in the Transvaal are getting on against Sekukuni, Chief of the Bapedi, who had lately defeated the Boer levies sent against him. This is evidently a tender point, and not at all appreciated, especially as our Kaffirs snigger considerably when this topic is introduced. They, too, have, by some mysterious telegraphy, known only to the tribes, heard of Sekukuni's successes in his distant mountains, and there is no love lost between Boer and Kaffir, chiefly owing to the bad treatment invariably practised by the former to the latter.

But time is getting on and it is four o'clock, and at half-past five we have an athletic event coming off, which has aroused all the sporting instincts of the immediate district. Not far from us is a Kaffir kraal, and amongst its inhabitants one young man, Segani, is renowned for his running powers. One of our own party, recently arrived in the Colony, has also acquired a reputation in England as an athlete, and for the last few years has been, amongst amateurs, at the top of the tree at distance-running, more especially from half-a-mile to two miles. The question arose whether the hardy Kaffir, inured to mountain exercise and capable of running immense distances, would be a match for the young Englishman, who had proved himself able to run a mile in four minutes twenty-eight seconds at Lillie Bridge, and had gained many a prize on running path and green sward. The Kaffirs were decidedly of opinion that their champion could not be defeated, while our party were equally confident of success.

Thus a friendly match was made for a race of

one mile, on the best part of the road near us, which was pretty fair going for South Africa.

A mile had been measured out, and already the Kaffirs could be discerned about half-a-mile away, near the winning-post, discussing and gesticulating in excited groups.

Our champion having donned the nearest approach to running costume he can find, and assumed a pair of canvas shoes which, luckily, he has by him, is ready. Proper spiked running shoes he has not, and indeed, on this occasion, they would be of little use; for though it is the best we can find, the road is "far, far from gay," and has a fair number of stones upon its surface, and to spring with spiked shoes upon these would be, to say the least, unpleasant. John, our Kaffir groom, who looks after the horses, slaughters our daily goat, and does odd jobs round the house, is anxiously waiting to accompany us to the scene of action. John, it may be mentioned, is a tall, wiry Gcaleka, standing six feet two inches, and has not long since arrived here in search of work from the territory beyond the Kei, commonly known as Kaffirland. He is not a bad sort of fellow if he is rubbed the right way, but, like most other Kaffirs, he considers himself a gentleman (and so, indeed, many of them are), and likes to do his work in his own way, and, it must be admitted, at his own time. To-day, John is resplendent in a complete suit of coat and trousers, and, moreover, wears above his good-humoured face an old top hat, recently purchased, in which he, no doubt, expects to create an impression on the Kaffirs who dwell in the adjacent kraal.

All being ready, we stroll down to the winning-post, where we meet the chief of the Kaffirs of the district, who resides at the kraal before mentioned. About forty other Kaffirs, the Boers recently arrived, with a few Hottentots and Bushmen, make up the "gallery." The baboons on the mountain-side yonder, the mountain eagle hovering motionless near the cliff tops, and the vultures soaring far away, mere specks in the pale blue, evening sky, who doubtless note our every movement, are the only other spectators of this curious scene.

The Kaffirs are chattering at a great rate, and a betting man, if he were present, would have no difficulty in laying a good book against their pet. The dark champion is not attired in the costume sacred to Lillie Bridge or Fenner's Ground; to speak the truth his costume is very nearly *nil*. He looks, as he always does, in hard condition, and his brown skin bears that peculiar and splendid polish that betokens in the Kaffir the best of health.

Three watches have been set at the same time; one of these goes with the starter to the starting-post, another is held half-way, and the third remains at the winning-post. These precautions are adopted as we are rather curious to know what a Kaffir really *can* do in the way of running. The respective champions and starter then walk quietly along the road till they reach the starting-point; they are unable to carry on the light and airy conversation affected by our English athletes on such an occasion, which but too often ill conceals anxiety and nervousness within; for, to most persons, the few minutes of preparation immediately before a race are very trying moments indeed. They are accompanied part of the way by about half the onlookers, who stop at the half-mile mark, where there is a slight bend in the road, and from whence they can witness the start. It was agreed that the race should be started by a rifle-shot fired behind the runners. The Englishman has doffed his hat and coat, and now stands ready in a light jersey and a pair of old flannels cut down to just above the knee. He is not so fit as he would be for an English championship, for he has not done much running lately; but for the last month he has had any amount of hard walking exercise while shooting about the mountains, and is in excellent health.

The contrast between the rich chocolate colour of the Kaffir and the white skin of the Englishman, as they stand prepared, is striking; as to physique, there is not much to choose between them. Segani stands about five feet eleven inches, weighs ten stone seven pounds, and has splendid muscular development about his chest and shoulders, and the dying sun, lighting upon his gleaming skin, shows up these features to perfection. Below, he is not quite so well set up as his opponent, nor are his legs and thighs so good; but he is an excellent specimen of a Kaffir, and his friends can put forth no stouter champion. The Englishman stands six feet one inch, weighs, in training, eleven stone; though slim and somewhat lean, he has broad shoulders, with plenty of room for lung power, to which, no doubt, he owes his staying qualities, and is well set up on excellent clean-cut limbs.

The brother of our host takes up his position behind the two whom he has placed on the starting scratch. "Get ready!" he calls out; another second, and the rifle cracks.

Let us watch now what tactics will be pursued by the runners.

Segani jumps off with the lead at a fair pace, but to his opponent, who has made up his mind to run a waiting race, it is but moderate, for he has been accustomed to be taken along for the first quarter of a mile in sixty seconds and less.

The styles of the two men are widely different. The Kaffir holds his arms somewhat low, and runs with a short stride, yet he moves easily and well. The Englishman strides quietly some two yards in the rear, with a free, machine-like action, and makes good use of his arms. So they journey for a quarter of a mile, when the Kaffir takes a glance round, and seeing his adversary just upon his heels, quickens up a little, evidently not quite satisfied with the way things are going. However, the second man shows no sign of coming up to him until they reach the half-mile (time two minutes thirty seconds), when the Englishman goes to the Kaffir's shoulder, and runs stride for stride with him for a dozen yards, just as a feeler. Segani is not quite happy, but he increases his pace, and the Englishman again drops back, with just the shadow of a smile upon his face. He knows now exactly what he can do with his man.

Both men, as they approach three-quarters of the journey, can hear their partisans shouting in the distance; some of the Kaffirs follow at their sides cheering on their champion. Ever since the half-mile post, Segani has been trying to get away from his man, and the Englishman in turn has been pushing him hard every foot of the way. Look at them now; the Kaffir is in trouble, his anxious face and laboured breath proclaim it but too plainly; he has been used to trot all day about his native mountains in his own way, but to be pushed hard just beyond his pace in this fashion is a new sensation to him, and evidently he doesn't much like it. The Englishman is striding along as freely as when he started, still a few feet in the rear. Three hundred yards from home he suddenly goes to the Kaffir's shoulder, and the real struggle begins; for thirty yards they race side by side, but it is soon over. The Kaffir is beaten; he is breathing hard, and his legs feel like bars of lead; and our champion, drawing right away, and finishing at sprinting pace, has won the race by seventy good yards, which, if he had made his own running, he might have increased to double the distance.

The mile has been run in five minutes, two seconds, not bad time considering the state of the road, which in many places is of a soft, sandy nature, and utterly unlike an English highway.

The delighted English crowd round their champion, eager to pat him on the back for having so well defended the honour of the Old Country. As for the Boers, who have stood in blank amazement at the whole proceeding, they ejaculate, "Allemaghte! vat zoorten mensch ist de?" (Almighty! what sort of man is this?) and evidently cannot at all understand why two people should take all this trouble for what seems to them no profit. The Kaffirs, rather cast down, gather about Segani, to know how his defeat happened; the poor fellow has done all he knew, and in reality has made a very fair display, and was simply beaten by a better man. After the race, our host hints to the chief, that Segani's conqueror is perhaps a better man than he was taken for, and the chief seems to be no little surprised when he hears of the white man's achievements. It is now getting dusk, twilight is of the briefest, and the warm red light in which the valley is bathed will soon disappear; so we at once return to the farmhouse. Our host has invited the chief, Segani, and one or two others, to come up and have a "soupje," and something to eat; and, needless to say, the offer is accepted with pleasure. In a short time the Kaffirs are being regaled at the back of the house, and, from their laughter, have evidently fully recovered their spirits, Segani's voice being the loudest of all. Presently they retire, after a gift of tobacco, and after Segani has been presented by his whilom opponent with half-a-crown, at which the Kaffir's face beams with delight.

Then we re-enter the house to finish our own suppers, after which, not the least pleasant of many happy days in South Africa is completed over pipes and grog, by an evening's chat and stories of shooting, and of old days in the Colony. And so, as good old Pepys hath it, to bed!

Chapter VII.

VAAL RHEBOK SHOOTING.

IN the long and brilliant array of South African antelopes, the vaal or grey rhebok (*Pelea capreola*) has, like its fellow mountain dweller, the klipspringer, for some reason or other, been unaccountably neglected by hunters and naturalists. Widely scattered though it is over nearly every part of South Africa, affording as it does excellent hill shooting, and possessing certain curious and indeed unique characteristics amongst the antelopes, it has yet been too frequently passed by unnoticed, in the rush to describe the larger and nobler game of this game-abounding country. From this unappreciative category, however, Gordon Cumming must be excepted, for, in his earlier days in the Colony, before he passed on to the vast and then comparatively virgin hunting grounds of his beloved limpopo and the far interior, he describes the vaal rhebok as affording him the nearest approach to Highland deer stalking of any game he had encountered. Cornwallis Harris, too, in his magnificent book, descriptive of South African game, has a good word to say for this shy and active antelope. Since the year 1836, when the emigrant Boers, discontented with British rule, shook off the dust from their feet, and trekked into the then unknown wilds of the territories now known as the Orange Free State and Transvaal, the tide of colonisation has too often swept into those countries, leaving the Cape Colony unnoticed and uncared for. The plains of the Free State, and the high veldt of the Transvaal, which when discovered were literally overburdened with almost every description of game, provided easier and more abundant shooting than the older Colony ; and, in consequence, the game in those States has been so harried, so depleted and exterminated, that there is now in the Cape Colony itself better shooting than in many parts of those sadly-abused hunting grounds.

The Cape Colony is an easy-going and a non-assertive country, and has been so long accustomed to hear of the brilliant shooting amongst the big game of the far interior, that its own merits in the matter of sport have been left unnoticed and unsung. And yet it can still boast, amid some of its denser bush-veldt country, and in the Knysna Forest of the elephant, the buffalo, and the koodoo; while, as I have shown in a former chapter, the noble zebra still adorns its mountains, in which the leopard also is common. Many travellers, passing up country, dismiss the old Colony with a sneer, or, perhaps, admit that there are a few springbok on the flats. These people, who follow the beaten tracks, and stay only at the accommodation-houses, know nothing of the deep kloofs, the wild mountains, and untraversed karroos of many parts of the Colony—places almost as primeval, as secluded, and as little known even now as they were fifty years since. To those who like a wild rough stalk amidst magnificent mountain solitudes and sombre kloofs, that even to this day know not the distant bleat of the farmer's flocks, the chase of the rhebok offers many attractions. In such localities, and especially in the zuur veldt, or sour-grass country, the klipspringer, the vaal and rooi rhebok, and the duyker may be found in plenty. In bygone days, when the plains of Cape Colony abounded with antelopes and other game, I can understand that the timid and retiring rhebok, with its modest garb, contrasting poorly with the larger and more gaily painted of the antelopes (such as the gemsbok, the eland, the hartebeest, koodoo, and others), was left undisturbed in its mountains; but at the present time, hunters cannot pick and choose so freely, and this antelope, I repeat, can offer very excellent sport after its kind.

Like the klipspringer, of which I have written previously, I first saw the grey rhebok near Naroekas Poort. A few days after arriving there, I rode out with my two travelling companions and our host to a wide shallow kloof, some six miles distant, where the latter had fixed the headquarters of his breeding establishment for horses and mules. In this kloof, and upon the mountains around, large numbers of these animals ran in almost pristine freedom, and a little Boer named Tobias Verwey, who lived there with his wife in a small house, looked after them, and kept down the leopards—very troublesome neighbours—by the aid of strychnine, the only effectual method of dealing with these nocturnal marauders. On this day we rode first to Tobias's house, with two zinc buckets carried by a Kaffir. Now, little Tobias (for, unlike most Boers, he was a little man), who was, I think, the most genial and friendly Dutch Afrikander I ever met with, had a peculiar talent, common also to Kaffirs, Bushmen and Hottentots, for finding wild honey; and as there was an abundance of this delicacy in the rocks around, he was directed, with the Kaffir, to fill the buckets—a business, including the finding, of little more than half-an-hour's labour—while we others rode a mile or two farther into the mountains. Accordingly we continued our way for a couple of miles, until we reached a part of the kloof resembling, more nearly than anything I can think of, the commencement of the wild pass of Glen Lyon in Perthshire, where the lofty frowning hills on the right-hand side slope steeply down to the river beneath. Just as we entered upon this grand bit of scenery, having emerged from a small mimosa grove that fringed a dried-up river-bed between the mountains, our host, with the word "rhebok" on his lips, pointed suddenly to his left; and less than a hundred yards distant a troop of six grey antelopes, rather taller and slimmer-looking than fallow deer, after eyeing us for a second, bounded away up the rough and precipitous mountain-side as only rhebok and klipspringer can. On this day, as we had not come out on shooting intent, only one of our number, one of my companions, carried a rifle, and he, unluckily, had lagged a little in the rear. By the time we had hastily called him up, the rhebok were quite 300 yards away, and although he quickly dismounted and had three or four long shots, the antelopes escaped untouched. It was marvellous with what free smooth strides they sailed away obliquely up the long stretch of mountain, hereabouts covered with boulders, loose stones, and low brush, and I suppose some 2,000 feet high, until at length their diminishing forms vanished over the mountain tops. Sorry as we were to have missed obtaining a specimen and some venison, we were yet well pleased to have set eyes upon a curious and, to us, new species of the antelope tribe. Our host having shortly finished inspecting his mountain veldt in this locality, we turned our horses' heads, and, after an hour's absence, reached Tobias's house again. Tobias, his expressive black eyes beaming with pride (I think the little man, from his dark complexion and low stature, must have been, upon his mother's side, like many other Boers, of Huguenot extraction), was ready for us with the two pails, each nearly filled with rich dark honey. After looking at a few horses in the kraal near the house, and learning that all was well with such of the stud as we had not seen on our way hither, we set off homewards, leaving Tobias and the Kaffir to follow with the honey. Here I may remark that the wild honey found in such abundance in these mountains, and indeed all over South Africa, though dark in colour, is as delicious and well-flavoured as any I ever met with, and, as may be imagined, it formed a staple luxury in the somewhat attenuated list of our up-country fare.

It was not long after this that I made closer acquaintance with the rhebok. One bright, clear morning, a few days later, I started away from our lonely farmhouse in Naroekas, accompanied only by Igneese, the Kaffir hunter of whom I have previously made mention, to shoot over a part of the mountains six or seven miles distant. The rest of the party were either busied in another direction, or, after a hard week's tramping about the hills after klipspringer, duyker, and feathered game, were inclined for a rest. I carried a Cogswell and

Harrison ·450 sporting rifle, while the Kaffir shouldered a mixed weapon, of which one barrel was rifled while the other carried shot. This Kaffir, Igneese, was one of the best native specimens I ever met with; a good shot, an untiring walker; and with powers of tracking and finding game unequalled, he lacked the noisy garrulity of some of his brethren. In addition to these good qualities, he was a careful, steady man, and possessed a good flock of goats and some oxen of his own, and was, in fine, quite a model Kaffir. We made an early start, and, descending the hillside on which our farmhouse stood, soon reached a path that led through the grove of thorny mimosa (*Acacia horrida*) along the bottom of the poort. We followed this path for a couple of miles, until we arrived at a point where two kloofs opened up into the mountains. Choosing the right-hand path, we plunged into a lonely gorge that led for miles into the heart of these rugged hills. The kloof was of surpassing beauty. Above and around us the mountains, clothed here and there with dense bush and scrub, reared themselves on either side to a great height, their dark brown crests showing strikingly against the clear azure of the sky. A tiny stream of pure water flowed by, while flowers innumerable spangled the bottom of the valley. Geraniums and pelargoniums flowered in wild beauty and luxuriance; many kinds of beautiful heaths and irids pied the ground with brilliant hues; aloes held aloft their rich red spikes of flower; whilst the Kaffir apple, Hottentot cherry (aasvogels besjes, or vulture berries, as the Boers call it), wilde pruimen (wild plum), euphorbia, prickly pear, and many another shrub, blossomed on every side. Hundreds of ringdoves cooed softly around, while gorgeous honeybirds, finches, and kingfishers, flitted from bush to bush, or darted arrow-like down the stream. The little rock rabbits (the coney of Scripture) glided merrily about the surface of the rocks, and here and there big rock pigeons bustled on fleet pinions. The soft and quiet beauty of this kloof, on that calm, warm morning, teeming as it did with the life of birds and the luxuriance of flowers, was a thing not to be easily forgotten. On such a day, and amid such surroundings, the rougher and ruder phases of South African life up country—and they are not few—are easily forgotten and forgiven.

Presently, as we cross the sand bed of a dry water-course, the Kaffir stops, and points to the fresh spoor of a leopard, but explains that the night marauder has long since retired to his mountain den. A few years before, when my host first settled in the poort, these animals were very numerous, and committed great havoc amongst his colts and flocks. From their nocturnal habits, they are very rarely seen in daylight, so that war had to be waged against them by the use of strychnine concealed in meat; they are now somewhat less numerous and less troublesome. Presently we strike again to the right, and ascend the mountain; it is stiff work, indeed, and requires both lungs and muscles to be in good trim, which, happily, is the case with both of us. Having traversed another kloof, and ascended another hillside, we near the ground where we expect to find rhebok. After half-an-hour's further very severe climbing, during which time Igneese was anxiously scanning the mountain-sides, and searching here and there for spoor, we came upon fresh indications of our game. Keeping what wind there was well in our faces, we now hastened cautiously forward, until in another mile we reached a shoulder of the mountain-side. Here Igneese again hit off the spoor, and very quietly we crept through some bushes till we could obtain a better view of the hill before us. At length, straight in front of us, uphill some 700 yards distant, we could see five rhebok, their grey forms showing so indistinctly from the rocks amongst which they grazed, that for some time I looked in vain. The ground directly between us and them was so open, that it was impossible to approach more closely from our present position; there was, therefore, no help for it but to creep back, ascend the mountain rather higher up, where the covert lay better, and then stalk in as well as we could.

After an exceedingly long and careful stalk, and by dint of crawling some 200 yards on our stomachs, we had managed to creep to within 140 yards of the herd. To approach nearer, without disturbing the antelopes, was impossible; so, after waiting a short time, in the vain hope that they might graze towards us, and after motioning to one another which side to shoot, we took steady aim and fired.

The rhebok I had selected half fell to the shot, but was quickly up again and struggling after the rest uphill; the Kaffir had missed clean, much to his chagrin. The wounded buck only ran for about a quarter of a mile, and we were luckily able to keep it in view, and in a very short time Igneese's knife put an end to its dying struggles. My bullet had smashed the angle of its shoulder, and, after passing

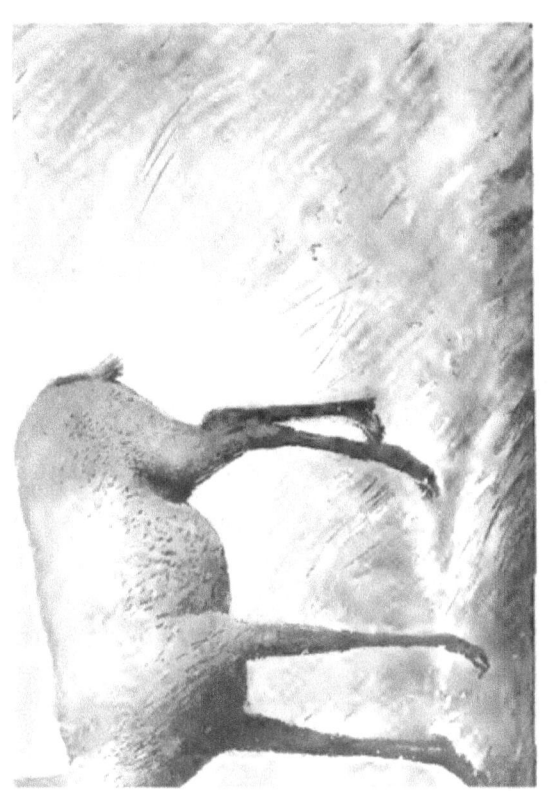

THE VAAL RHEBOK

nearly through the body, had run off at a tangent up the neck nearly out at the skin on the off-side. The bullet, hollowed at the top of the cone, had caused a severe tearing wound, and the wonder seemed how the antelope could have moved at all after such a shock. To people accustomed to African shooting, however, the extraordinary vitality of the antelope tribe, and, indeed, of every kind of game, is familiar; and even when hit in apparently a vital spot, the game will over and over again escape from the hunter. One reason for this, as it seems to me, is that in these parched and rugged wastes Nature has provided an increased store of strength and endurance, to enable her offspring to contend with the many difficulties and dangers that they have to undergo, whether they be attributable to waterless deserts, fierce animals, or the merciless Bushmen, Hottentots, and Kaffirs, who, through unknown ages, have preyed upon the game that surrounded them.

We were now in the centre of the wildest part of these mountains, and as the beauty of the prospect around could scarcely be improved upon, I rested a short while for lunch—some biltong and biscuit, and a mouthful of Cango brandy. While we rested, Igneese gralloched the rhebok, and I had time to notice its proportions. The vaal, or grey rhebok (*Redunca capreola* or *Pelea capreola*), Peeli of the Bechuanas and Kaffirs, is, as its name indicates, of a grey colour. It varies considerably from the rooi or red rhebok, which is also found in these mountains and in other grassy localities nearer the coast line, more especially towards the eastern part of the Colony. It stands at the shoulder some two feet six

inches in height, and is in length about five feet.
The body is light and slender, the ears are long
and pointed, head smallish, while the horns are
straight, thin, and pointed, about nine inches long, and
annulated at the base; the belly is white, or nearly
so. The female of this species is lacking in horns,
and is smaller than the male. The most peculiar
feature of this antelope lies in its coat, which differs
essentially from every other South African antelope,
and consists of a thick woolly fur, approaching very
closely to the texture of that of the rabbit, but softer,
finer, and longer; the venison, although inferior to
springbok and some of the larger antelopes, is by no
means despicable, but it has the fault, common
to much South African game, of being somewhat
dry. Although as regards beauty this antelope may
not be compared with many of its congeners, it has
its good points; when extended, it is a most
beautiful mover, and the nature of its habitat and its
shy habits will always render it worthy of the
sportsman's skill. I believe I am correct in stating
that there is no specimen of the rhebok in our
Zoological Gardens, nor am I aware that any
specimen has ever been shown alive in this country.
I do not think the vaal rhebok runs in troops of more
than eight or ten at the outside. I think eight was
the largest number I ever saw together; oftener from
four to six were seen. The spot where we rested,
high up among these rude hills, was indeed desolate.
The foot of man but seldom treads these remote
regions; a hunter now and again, or a few farmers,
perhaps once in a dozen or even twenty years,
beaconing off their boundaries, are the only human
beings likely to disturb the repose that reigns so fitly

here. This dead antelope has a certain charm in my eyes. It is the representative of the matchless fauna that imparted life to these solitudes long ages before Van Riebeek and his sturdy Dutchmen set foot in Table Bay; long even before the little Bushmen hunters possessed the land, or the Hottentots—the Khooi-Khooin (men of men), as they proudly call themselves—held sway, or the Kaffirs, pushing their way southward and westward through the dark continent, at length appeared upon the scene.

My musings were soon over; we shortly resumed our march, and tramped on for another hour without seeing further traces of game, except a brace of klipspringers, far out of range, and some francolins, which we decided to leave in peace. At length, after laboriously ascending and descending alternate mountains and kloofs, whose crumbling and jagged sides began to make us weary, we suddenly came upon a small troop of four rhebok, which had either got our wind shortly before, or had heard our approach through a kind of *nek* or dip in the hills; they were slowly cantering off below, and directly in front of us, their long ears and somewhat stilty legs making them look larger than they really were. The distance was considerable—about 400 yards—but Igneese and I both let fly at the retreating herd. As I expected, I made a clean miss, but the Kaffir was more fortunate, and wounded a bok on the left hand. It was a difficult shot, and down a steep mountain-side. Reloading as we ran, we followed the now flying antelopes as hard as we could tear; but, clearing the rough ground as they did, apparently with the ease and smoothness of a racehorse at exercise, we stood no sort of chance with our fleet quarry, and they were soon far beyond ken. The bok wounded by Igneese must have been very slightly hit, for after searching and spooring about for another hour, we were obliged to abandon pursuit, and turned homewards, as it was now getting towards three o'clock. Another *soupje* from my flask, and a well-earned pipe of tobacco, and we were striding up hill and down dale again until we had reached the rhebok shot in the morning. It was too far home to carry the whole carcass, so Igneese quickly skinned the animal, and taking the best of the meat, hid the remainder, which he intended returning for early next morning.

The Kaffir having taken as much meat as he could carry, we started homewards, having still a longish march before us. As we descended a craggy steep some way farther on, Igneese suddenly touched me, and pointed to a deep thickly bushed hollow in the kloof right below us; I shortly made out a dark form, half hidden by the brush some 300 yards beneath. The duyker, for such it was, had for a wonder not noticed our approach, and I had time for a steady shot. What with the distance, the bad mark offered—for the bok was more than half hidden—and the difficulty of shooting almost perpendicularly down hill, I missed, however. This antelope (*Cephalopus mergens*) is common in the Cape Colony, and may be found in bushy cover, and especially in such deep dark bottoms as I have described. It stands about two feet high, is of a dark brown colour (white underneath), carries short, straight, sharp horns, four inches long, and is found either singly or in pairs. Its Dutch name, duyker— that is, diver or ducker—aptly describes it, for it invariably dives or sneaks into the densest cover it can find, and will even lie down in the hope of escaping notice. The duyker is not a favourite with most Cape farmers, who, I fancy, despise it for its sneaking characteristics.

Again resuming our way, we proceeded supperwards. By the aid of the shot barrel of the weapon which Igneese had carried, I knocked over a brace of fine rock pigeons in the flower-spangled kloof we had passed through early in the morning, and, just before we reached Naroekas, a spring haas, or jumping hare—one of the jerboa family—which was hopping along in our front in the dusk of evening. Our bag was not a great one, but the grand and diversified mountain scenery, the pure air and long rambling walk, and the sight of the rhebok and other game, had amply repaid me for such fatigue as had been undergone.

CHAPTER VIII.

A SPORTING SAUNTER.

MY ramble to-day is a solitary one, for my friends are busied in other directions around our mountain farm. Shouldering only a shot gun, and leaving perforce behind me our two sporting dogs—a retriever and setter, which are only just recovering from the dire effects of eating poisoned meat intended for leopards—I saunter down the boulder-strewn mountain-side, and direct my footsteps by a dry stream-bed through a little forest of thorny mimosa. The Cape winter has passed away, and the rich and diversified flower-life of this often-maligned land is approaching its best and bravest. The yellow acacia bloom, fragrant with sweet scent, is beautiful to look upon, and contrasts strikingly with the shrivelled winter appearance of these trees. Amid the branches, hundreds of ring-doves—the *tortel duif* of the Boers—coo softly and soothingly in the pleasant warmth. After a walk of two miles, I approach the sharp angle of a mountain spur, whence two kloofs or gorges run for miles into the hills. Choosing that upon the left hand, as being the wilder and more solitary, I move quietly under the sheer rocks, which here run upwards like a mighty wall. Just as I turn the corner, where I know the great rock pigeons love to rest, I look upwards, and surely enough, thirty yards above, on a rocky ledge, sit a brace of these handsome birds; they have not yet perceived me, but directly I betray myself, they bustle off on strong and noisy pinions. My gun is up, and with the report, the nearer bird turns over in mid-air, and falls heavily to earth, its fellow somehow escaping the second barrel, much to my disgust. Below the rocky cliffs, just where the pigeon fell, the wild geranium, or pelargonium, grows in bewildering luxuriance, and, being retrieverless, I have some difficulty, wading through the flowers, thick and middle high, before I pick up my spoil; at length it is secured and I resume my march. The kloof lying before me is here 200 yards wide, and is shut in on either hand by dark brown frowning hills, their sides here and there lit up by ruddy splashes of lichen and the blood-red flower spikes of the aloe. A stream of clear water, down which I note the swift flash—mazarin-blue and red—of a kingfisher, runs through. The rich bottom soil is gay with flowering shrubs and bushes, and—besides the pelargoniums, brilliant heaths, ixias, gladioli, and other irids—amaryllids, orchids, and other flowers, bewildering in their splendour and their plenty, star the ground.

The flora of the Cape, perhaps the richest in the world, seems created in truth but to blush unseen. At home it is almost unknown, yet English botanists, if placed in a South African mountain kloof, or upon the boundless karroo, would give their eyes and ears for the beauty around them at certain seasons. Proceeding quietly, in a few hundred yards, I come upon four or five more pigeons, sunning themselves upon a rocky krantz; my powder this time proves straighter, and a neat right and left tumbles over a brace of the gallant birds. The improvement in my shooting adds just the required fillip of zest to the charms of the scene around, and with elastic step I move onwards more rapidly towards the ground where I expect to find the red-wing partridges. There is a track plainly visible as I walk, though it is here and there overgrown by the luxuriant vegetation —a track worn by centuries of the ponderous tramp of elephants that once browsed amid these mountains. But the elephants have been driven, more than a generation ago, by the hungry ivory hunters, to their last strongholds in the impenetrable bush-veldt of the Sunday and Great Fish Rivers, and the dark jungles of the Knysna Forest, where they yet linger in the Old Colony. For the space of nearly half-an-hour there is no game to be seen worth the shooting, although I am not alone. In the bushes near me the sun-birds or sugar-birds— *Les Sucriers* of the old French traveller, Le Vaillant (who, by-the-bye, crossed these very mountains in 1784, through a poort not twenty miles distant), flutter restlessly from flower to flower, extracting the sweet juices with their long tongues. Songless though they are, the plumage of these birds is glorious; brilliant shining greens, marvellous blues, reds, and yellows, and orange gold predominate in their colouring. Solomon, in all his glory, was not more gorgeously arrayed. At length I espy a bird long wished for—a bird that creeps slyly in and out of the brake, as if to escape the eyes even of its feathered fellows. I know him at once for the bush lory, and hastily slipping in cartridges of finer shot, I quietly follow him up, till, as he emerges from a bush, I secure him with the first barrel. This is a prize of value, and after examination and admiration, is tenderly bestowed in an inner pocket. The bush lory—*Le Coucouron Narina* of Le Vaillant, who named it after the fair Gonaqua Hottentot, Narina, of whom he writes so gushingly—with its vivid green back, grey wings, delicately pencilled in white, carmine breast, and green throat, is a curious and a notable bird. Having bagged the lory, I hasten on, and in another half-mile I turn an abrupt corner, or hoek as the Boers call it, of the kloof, and closely and cautiously scanning the cliff sides, I set eyes upon three klipspringers—literally, rock jumpers—standing upon a ledge half-way up the mountain-side. The klipspringer may be styled the chamois of South Africa, and is one of the handsomest, most characteristic, and shyest of the antelopes of the country. One of the trio is but half-grown, and in size about equals a smallish lamb. For ten minutes, hidden in the bushes, I watch the beautiful, sturdy little creatures, leaping here and there, from rock to rock, and ledge to ledge, feeding as they go. They are not very far away—although out of range of my shot gun—and I can plainly distinguish the rich olive brown of their coats, and the glint of the sun upon the short straight horns, sharp as poignards, of the full-grown ram.

A Kaffir hunter of ours saw near this very spot a singular scene. Here is his story:—" I had been hunting all the morning, and at last came upon two klipspringers, on the mountain-side, perched upon a jutting-out rock. Just as I was about to creep closer for my shot, I noticed another hunter, keener even than myself. Above the klipspringers, just over the cliff-tops, and unseen to the antelopes, hovered a great black eagle, a berghaan (*Le caffre* of Le Vaillant, the dassie vanger—rock-rabbit eater—of the Dutch), and he, no doubt, had long had his piercing gaze upon their movements. Just as I was about to move forward, one of the antelopes leaped down to a tiny pinnacle, or peak of rock, jutting still farther from the mountain-side, and at that instant the black eagle swooped like lightning from above, and before I could realize it, or the poor klipspringer either, with talon and beak, and the rush and force of its mighty wings, had swept the hapless antelope from its perch headlong to the rocks beneath. Black as death was the eagle, and deathlike was its swoop. This was more than hunter's flesh and blood could stand, and rushing forward, I was in time to drive the eagle from its quarry, which lay struggling with broken back and smashed head. Quickly I put the poor little buck out of its misery, and hid it away under thorns and stones, until I had finished my day's hunt." Such incidents are not uncommon.

A mile or so farther on the kloof broadens out, the mountains slope less steeply, and the bottom of the valley rises upwards somewhat. Here I must " gang warily," for the " red wings " abound on the more open ground. The wind is in my face, luckily. I still hold up stream, and presently, just as I expected, in a clump of palmiet that grows thickly in the moist ground, I flush a brace of big greyish-brown birds, that whirr away with even stronger flight than our English partridge. At twenty-five yards I stop one, but its fellow, wheeling at the report sharp to the left, escapes a too hurried second barrel.

Quickly I reload, for I know that these birds often lie thick and close in the palmiet, and ere I reach the stricken bird, four others rise almost from under my feet. Out of this quartette I secure a brace greatly to my contentment. It takes me ten minutes to gather up my spoil, for one of the three is only wounded, and I am not so handy at retrieving as an English spaniel.

The red wing partridge of the Cape colonists is, in truth, a francolin—*Francolinus Le Vaillantii*—and like many another South African bird, was discovered and named by the indefatigable if sometimes inaccurate Le Vaillant. It is a handsome game bird, greyish buff upon the top, dark orange red on the throat, and upon the speckled rufous brown chest there is a curious white half-moon, or gorget, somewhat similar to the ring of our English ring-ousel; there are also white lines upon the head. The insides of the wing-feathers are of a deep rufous colour, whence the bird takes its name. Whether considered from the ornithologist's point of beauty, or for its virtues as a table-bird, this francolin is in truth a prize well worthy of capture. A walk of an hour or so up the valley adds three brace further to my score, and I turn homewards well satisfied with my varied if modest bag. Although the Cape Colony is now nearly denuded of its larger game, save only the elephant, buffalo, zebra, koodoo, and leopard, which still linger in remote districts, Englishmen in search of change will yet find there an abundance of the smaller antelopes and a vast store of feathered game, and by moving from one farmhouse to another, any well-behaved persons will procure an abundance of sport at very little expense. They will behold a magnificent country, as yet scarcely touched by agriculture—a country still in many parts as primitive and as little known as when the Dutch first landed on its shores—and they will find a climate unsurpassed, even if viewed from the stand-point of health alone.

But since the passing of stricter game preservation laws by the Cape Parliament a year or so back, the close time for antelope shooting is extended to the months between July and January in some districts, and between August and March in others; and further, in some divisions certain antelopes are protected from the gunner altogether for periods ranging from one year to three. I have referred to this with more particularity at the end of the chapter on the antelopes and larger game of Cape Colony.

Coming homewards by the river, I disturb some blue cranes, which rise and fly off to some quieter resort; these dark-hued birds are not uncommon hereabouts. The great wattled crane is also met with, but only very occasionally. Sometimes wandering down the river very quietly, we happen upon these blue or Stanley cranes in a small band. When undisturbed and completely at their ease, for they are shy creatures, their pacings, dancings, marchings, counter-marchings, and other playful manœuvres (I suppose these movements *are* playful, although they are often gone through with awful solemnity) are indescribably ludicrous. Watching a group of cranes pursuing their pastime on a smooth spit of river-sand, I have sat for nearly half-an-hour, with one of my companions, in fits of smothered laughter at these absurd avi-faunal junkettings and philanderings.

Another long-legged bird—the violet stork—is very occasionally to be seen in this neighbourhood. This bird, the *Ardea nigra* of Linnæus, is much more remarkable in colouring than the slate-coloured, blue-black crane I have mentioned. A specimen of this fine stork was shot at Riet Fontien far out on the dry karroo, strangely enough; and its skin, which I brought home for a brother of the shooter, was very beautiful in its strangely metallic sheen. The general colour of the violet stork is a dark shining green-brown; the metallic lustre is very remarkable; the under parts and thighs are snow-white; the bill, which this bird uses very freely in its defence, is red, as also are the long legs. Before reaching home, I shot near the water a blue gallinule (*Porphyrio erythropus*), a handsome green and purple-hued bird, having legs and bill of brightest red. This member of the family of rallidæ (rails), is to be found near rivers and valleys all over the Colony. Finally, I gain the cool stoep of the farmhouse—now pleasantly shaded from the sun—well content with my quiet ramble in search of sport, and my modest bag, and in a fit state for war upon the supper, not long hence to appear.

CHAPTER IX.

BIRDS OF PREY IN CAPE COLONY.

AFRICA, and especially South Africa, is peculiarly rich in birds of prey. Whether you traverse mountain, karroo, forest, bush-veldt, or deep and jungly kloofs, you will never be long out of sight of some member of the order raptores, be it vulture or eagle, buzzard, kite, falcon or owl. Of these raptorial birds it may be asserted, from a study of the most competent authorities, that some fifty or more species are to be found within the limits of the Cape Colony (*i.e.*, the old Cape Colony proper, bounded by the Orange River on the north and the Kei River on the east). These limits embrace a vast extent of country, much of it scarcely known to the colonists, and still more quite unexplored to this day by ornithologists, and it is, therefore, not at all improbable that some rare birds of prey may to this hour be existing in melancholy obscurity, so far as collectors are concerned. Mr. E. L. Layard, in his "Birds of South Africa," acknowledges that many species on the frontiers, especially to the eastward, may be yet undescribed, as he had not been able to visit these districts personally, or get anyone to collect there for him.

The raptorial birds of Cape Colony, exclusive of the secretary bird, may be divided as follows: Five species of vulture, two buzzards, twelve eagles, two kites, seven falcons, thirteen hawks, five harriers, and nine owls. This is a goodly array, indeed, and it is not surprising that the Cape farmers, whether Boers or British, are their sworn foes.

Naroekas Poort, as I have said, is one of the wildest and most solitary mountain ranges of the Colony.

In this wild spot many birds of prey abounded. Whenever we went out shooting or farming, if we failed to catch sight of any vultures, soaring or circling in graceful sweeps over some object beneath them, we might always count on seeing an eagle near the cliff tops, or a hawk or falcon on the lookout for food of some kind or another; nay, without stirring out of doors, we could often watch one of the fierce tribe (usually a jet-black dassie-vanger or a Senegal eagle) hanging motionless in air above the mountain-side facing us, or stooping occasionally, with swift flight, upon its prey. Thus opportunities of noticing these interesting, if somewhat inconvenient, birds (*i.e.*, to the farmer) were almost constantly before us.

The bearded vulture, "arend" of the Cape Dutch (*Gypaetus meridionalis*), was a frequent visitor from the mountains, and, I believe, actually bred in the locality. This great bird is an extremely dangerous neighbour for lambs and kids, and as our mountain flocks consisted entirely of goats, a pretty good lookout was generally kept for it. The flight of this bird is, for power, grace, and swiftness, unexampled; its ordinary passage through the air seems made without any apparent effort whatever.

The black vulture, zwart aasvogel of the Boers (*Otogyps auricularis*), was another of our neighbours, as well as the common vulture (*Gyps fulvus*), aasvogel of the colonists. The word aasvogel, by the way, literally means flesh or carrion bird. Both of these species are bred in our mountains, and if we were shooting and killed a klipspringer or rhebok, or other antelope, their presence was, in a very few minutes, made known. It is extraordinary how these birds will, on these occasions, suddenly start as if from space. The common aasvogel is much more numerous than the black aasvogel, and will generally get out of its way at a carrion banquet. We shot specimens of these birds, and also found them dead from poisoned meat set for leopards, but did not care greatly to handle the loathsome creatures.

The witte kraai (white crow), or Egyptian vulture (*Neophron percnopterus*), was pretty frequently seen; in the Western province this bird is rare. The Kaffirs say it nests in the mountains, but I never saw its eggs. This is a singular bird, with its dirty drab colouring, relieved only by a touch of yellow round the eyes, cheeks, chin, and throat, which are bare.

Of the buzzards, we occasionally saw the jackal vogel (buteo jackal). It is a handsome bird, brightly coloured in ruddy brown and black, with a touch of white upon the throat; it builds in bushes principally; its cry is often heard, and is sharp and strident; it is death on rats and such small game, and will sit for an hour patiently waiting for its opportunity.

The eagles, troublesome neighbours though they are for the pastoral farmer, have a noble something about them, that to some extent, small though I fear it may be with the colonists, condones their crimes.

Of these grand birds, I think the dassie-vanger (*i.e.*, rock-rabbit eater) of the Boers (*Aquila verreauxii*) was perhaps the most striking specimen in our valley. In colour it is jet black, with a patch of white upon the lower part of the back; the legs are thickly feathered, the toes bright yellow, and the claws black. Often were these bold birds to be seen sailing about the mountain tops, or hovering motionless near the cliff sides. Our mountains abounded with little rock-rabbits (das or dassie of the Dutch), and these animals are the favourite food of this eagle, hence its name. It is said, and I believe with perfect truth, that the dassie-vanger will on occasion beat or hurl the klipspringer antelope from the peaks and projections of the cliffs and rocks on which it has its habitat, afterwards feeding upon its dead body. The Kaffirs in Naroekas always affirmed this, and Boers and other white men have frequently told me the same thing, and from the nature and habits of the birds, which we frequently found near klipspringer ground, it is not difficult to believe the story. The dassie-vanger will certainly prey on kids, lambs, weakly goats, and sheep. A specimen of this eagle was shot by one of my friends near the carcass of a dead kid. I remember one hot afternoon resting after a long tramp beneath the shade of a rock, and watching one of these eagles manœuvring for a rock rabbit. These little animals are extraordinarily agile and watchful, and slip about the rocks with lizard-like rapidity. Long and patiently the great sable bird waited aloft; at length its opportunity came; swift as thought it made its stoop, successfully seized its victim, and with it soared away again to its nest. The dassie-vangers are not, however, invariably, or anything like it, the winners in these games of skill, and I have frequently seen them miss their stoop. The nest of this bird is made on the most inaccessible rocks and ledges, and at great altitudes; and I was too busily engaged in shooting and other matters to essay the break-neck pastime of searching for one.

The Senegal eagle, coo vogel of the Boers (*Aquila senegalla*), was another familiar eagle with us. In size it is nearly on a par with the last-named, though a trifle smaller, measuring in length about two feet eight inches, and is a determined and dreaded foe to young antelopes and weakly and young sheep or goats. We almost always saw these eagles when out shooting, and I believe that on the Karroo plains—where I have also seen them frequently—they constantly accompany hunters, and even pick up wounded game birds.

One of my friends shot a specimen of this eagle. In colour it is a rich rufous brown, darkening towards the tail and wings; the toes, cere, and irides are yellow, the legs amply feathered, and the beak black. Mr. Layard speaks of the tameness, when in confinement, of a specimen of this bird, afterwards sent by him to the Zoological Gardens, and of the probability, from its habits, of its being trained into a good hunting eagle. Undoubtedly, from the keen interest it takes in hunting parties, and its amenity to semi-domestication, this is possible; and the Senegal eagle might be developed into as useful a hunting eagle as the "bearcoot" of the Khirghiz Tartars. The "bearcoot" is employed in Asia for hunting deer, wolves, and foxes, and, if well trained, was, according to Pallas, valued at the price of two camels. It is not likely, however, in South Africa, where the colonists are only interested in destroying the aquiline race, that the interesting spectacle of a trained hunting eagle will ever be witnessed.

The crowned eagle (*Spizaetus coronatus*), very occasionally, was seen in Naroekas. It was pointed out to me on two occasions, but I only witnessed it in flight. In size it much resembles the dassie-vanger, but has a longer tail; it is of a light colour, especially underneath, the back being dark brown slightly marked with white. The crested eagle (*Spizaetus occipitalis*) was also an occasional visitant, coming, I think, over the mountains from the forest district of Knysna, which it seems to prefer. This is a dark brown bird, having a long crest of blackish-brown feathers about half a foot in length; the legs are well feathered, and are in striking contrast to the dark body, being of a pure white. I saw one specimen of this bird which had been shot sitting on a rock; it is considerably smaller than the crowned eagle. There are at least two fishing eagles in the Colony, *Haliaetus vocifer* and *Haliaetus leucogaster*. The commoner of these, the first-named, I once saw on the coast not far from Port Elizabeth, some distance beyond Baakens River. I believe it is abundant on the coast of the Knysna district. This eagle has a good deal of white about it, and is called "witte visch vanger" (white fish-eater) by the Cape Dutch.

Passing to the falcons, the lesser peregrine, spervel of the Dutch (*Falco minor*, or *Falco peregrinoides*), was pretty frequently seen and sometimes shot, as it is a great troubler of the poultry; it is almost identical with the European peregrine. Another falcon, which I took to be *Falco biarmicus*, but whose colonial name I never heard, was also seen occasionally. In our valley, not very far from the farmhouse, there was a large grove of mimosa trees, down near the bed of a periodical stream, and amongst these mimosas there were hundreds of turtle doves. Upon these hapless doves the two falcons I have named, as well as others of the raptores, did great execution, scattering every now and again the poor wretches with sudden swoop.

The smaller falcons were not uncommon. The hobby (*Hypotriorchis subbuteo*), with its bluish grey back, was sometimes observed, as also was the rooi valk (red hawk—*Tinnunculus ruficolus*), a pretty little falcon, resembling our own kestrel, very frequently to be seen hovering in mid-air, watching intently for birds, rats, mice, and other small game. Seen in flight, this bird looks as its name implies, principally rufous in colour, but the head, neck, and tail are of a bluish grey.

Of the kites, I believe only two have been properly recognised in the Cape Colony. I saw one near Naroekas on two different days, which I took to be *Milvus parasiticus*, but these birds are not frequently seen, so far as I could learn.

Of the hawks, one of the commonest and yet most interesting was the blue hawk, sometimes called the chanting falcon, the blaauw valk of the Dutch (*Melierax musicus*). This interesting bird has been surrounded by Le Vaillant with a somewhat fictitious glamour. He tells us that it sings regularly night and morning—and even in the night—for about an hour, at intervals of a minute; and further, what is certainly correct, that it is shy and difficult of approach, but can be shot while singing, when it seems much engrossed. I fear the old-world French traveller was drawing upon his imagination for the first statement, for I have never been able to learn that there is any such regularity or method in its singing as he pretends. I often saw this hawk in and near Naroekas Poort, and when crossing the karroo it was even more abundant, and I have heard its singular cry or song. This is a handsome bird, rather long in the leg, and slightly resembling the jerfalcon. It stands some two feet high, its upper parts are of a greyish brown, while underneath it is white, barred with brownish blue lines; the legs are red.

The chanting falcon preys on hares, rabbits, francolins, bustards, and other game birds, and even upon the paauw (*Otis kori*), on occasion. All the rapacious birds have shrieks and cries of their own, but the blaauw valk differs from its fellows in possessing the power of making sounds almost musical in their cadences; these sounds have been compared to the thrilling notes of musical glasses, which indeed they do distinctly resemble. I have heard this hawk's notes in the early morning, but was never able to convince myself that Le Vaillant was correct in stating that it sang regularly morning and evening. A small dark brown hawk, which, I think, must be *Accipiter minullus*, was very occasionally seen at Naroekas Poort. Le Vaillant states that this hawk hunted in the country between Gamtoos River and Kaffraria, and there, I think, he is undoubtedly correct, for Naroekas Poort lies only a few miles from the Gamtoos or Groote River, and this hawk is found in other parts of the Eastern province. It has become the fashion to sneer at Le Vaillant as a mere fabricator of absurd stories concerning birds and their habits; but we must remember that Le Vaillant did, at a time when the Cape was a dangerous and almost unexplored country, actually discover, and accurately record, very many of the birds of South Africa, birds which to this day are distinguished by his name; and for this work we ought to be duly grateful. The hawk known as *Melierax gabar*, was fairly common in our neighbourhood. This is a light brown hawk, the darker colour on the back changing to grey upon the throat, and white lower down; we always found it in particular in a deep bushy kloof or gorge where small birds were plentiful. The harriers were not very well represented in our mountains, but as they are generally found in low marshy spots, this is not to be wondered at. The only specimen I saw was one of the species called by Le Vaillant " le grenouillard " (*Circus ranivorus*). This specimen was shot near the river which ran through our valley, where I expect it was hunting for fish or frogs. There were several of the owl family in our neighbourhood, but the only members definitely distinguished were the large-eared owl (*Bubo capensis*), *Le moyen duc* of Le Vaillant, and the common white owl (*Strix affinis*).

Our principal day for noticing quietly such of the raptorial birds of the Colony as favoured us with their presence was Sunday, a day which, in the more remote farmhouses (where church is, as was the case with us, some sixty or seventy miles' distant), hangs rather heavily after private service is over. On hot afternoons, in the grim old poort, often have we helped to pass the time watching the great eagles wheeling about the cliffs, or the hawks and falcons plying their devastating careers.

CHAPTER X.

AN UNLUCKY DAY.

I QUITTED our mountain valley one morning, attended by Jackson, an English mason, who was staying for some time at the farmstead building cattle kraals, and generally making himself useful upon the estate. A handy man of this sort, especially if he be a blacksmith, and sober and diligent to boot, may reckon upon constant and lucrative employment in South Africa. Jackson was a good enough sort of man, and was desperately keen for a day's shooting now and then, an instinct possibly inherited from some remote poaching ancestor in the Old Country, or, still more probably, naturally implanted in his breast in common with so many denizens of these islands. But Jackson had been out of health for some time from fever and rheumatism, and was not in the best condition for a long tramp over rough mountain veldt. However, the man begged so hard that I consented, and we set out at eight o'clock in the morning, he armed with our "mixed" weapon carrying shot and bullet, while I took with me my favourite "George Daw" ·450 sporting rifle. Varying our usual hunting ground, we made up the steep mountain directly at the back of the farmhouse, and after a hard scramble of five-and-twenty minutes or so, reached the crest. Here we took a survey of the country—a survey embracing in its sweep nothing but mountain tops, some peaked, some serrated, others of the broad flat table shape so common to South Africa. It was a glorious morning, and the clear crisp air—for it was yet spring-time at the Cape—the blue vault of space, and the grand panorama of hill scenery billowing around us, all appealed strongly to that sense of the beautiful, which I suppose every man, in some degree, slight though it may be, possesses. Having settled our plan of action, and noticed how the wind lay—a matter of vital importance in dealing with the marvellously keen-scented game of South Africa, and indeed every other wild country—we descended the mountain on the other side, bearing right-handed for a *nek* or opening between the hills. Reaching this *nek* we looked to our weapons, and then stole cautiously along for some two hundred yards over rough ground, covered with long coarse grass, until we came to a turn of the mountain. Rounding this turn, we came suddenly upon a reddish-coloured antelope that was grazing apparently alone some sixty yards away.

The rooi or red rhebok, for such it was, was away like lightning, almost as soon as we set eyes on it; and although we fired simultaneously, we were not quick enough, and the antelope had gained the shelter of some dense bush, and we never looked upon it again. Misfortune number one! This was the only buck we got within range of throughout that day, although we sighted others. Doubtless, if we had had Igneese, our Kaffir hunter, he would have *somehow* got us within fair shot of the rhebok; for these people, from centuries of practice, have much finer instincts in tracking and finding game, than as a rule is to be found in the average white man;

however, nothing daunted, we resumed our way. All that morning we toiled up rough and jagged mountain-sides, through deep kloofs, and along flat, table-like, hill tops, littered here and there with huge dolmen-like boulders, and covered with dense crops of long waving sour-grass, in search of game; it was a hot morning as the sun gained strength. Twice only did we sight antelopes before twelve o'clock; once a brace of klipspringers, far up on a rocky krantz, and quite inaccessible, except at the long range of six or seven hundred yards. Lying flat upon the ground, we put up our sights, and essayed the forlorn hope. As I expected, the bullets struck short, for in the marvellous clearness of this atmosphere, one is apt too often to misjudge distances in this way; only incessant practice and the constant habit of pacing out distances can get over this difficulty. As the bullets rattled into the cliff below them, the klipspringers bounded away up the precipitous crags, and were soon lost to view; then the echoes of our shots, reverberating loudly among the rocks, came rolling back. We obtained just a glimpse of a vaal or grey rhebok a little later, but he got our wind, and was away long before we could get near him. At twelve o'clock we rested for a mouthful of lunch, and I then discovered that Jackson had had about enough of it; the lack of sport had no doubt something to do with this, added to his somewhat indifferent health. We determined to walk quietly for another hour or two, and then, if unsuccessful, to make our way to a store that lay just outside the poort, or pass from the mountains to the plains beyond. Still no luck; we tramped steadily on for two hours longer, and then rested upon a

TROOP OF OSTRICHES

mountain top that overlooked Waai Kraal (Windy Kraal), the store I have mentioned.

It was a fine prospect. On our right the Gamtoos (hereabouts more usually called the Groote) River, or the remains of it—for drought prevailed—meandered peacefully beneath, its shallow course occasionally broken into a chain of deep pools. Just on its thither bank nestled three or four small Boer homesteads, around which, upon some rich alluvial soil, grew pleasant fruit trees—oranges, peaches, grapes, and quinces—although at present not in the full bearing—save the oranges—of their summer fruit. Further to the left the white walls of the store shone, gleaming beneath the hot sunlight, and in a camp near the house we could just discern a few ostriches pacing solemnly hither and thither. Away in the distance, far as the eye could reach, the open Karroo plains stretched to the foot of some distant range that lay purple upon the horizon. As we rested here, the only bit of sport, which fell to us this day, happened. Jackson rose, and went poking about in some long grass with his gun; a solitary bird got up from under his feet, and letting drive with all the bottled-up eagerness of desperation, he knocked it over with his shot-barrel at fifteen paces. The victim proved to be one of the plover kind (*Charadrius coronatus*), and is known all over the Colony as the kiewit—a name evidently given to it, as in the case of our own peewit, from its piping call. Purplish drab as to its upper parts, its breast is of a darker shade of brown; the stomach is snow-white, and the tail feathers are also white, marked towards the extremities with black cross-bars, and tipped with white; the crown of the head is black, effectively separated by a white line. This bird is seldom found alone; in this instance its fellows must have been disturbed by our vicinity.

Picking up his solitary and battered prize, Jackson came back and finished his pipe. Although temporarily a little roused from his fatigue, I could see that the man's weary soul no longer hungered after sport; in fact, he was now completely knocked up, and our only plan was to make for the store, rest an hour or so, and then, towards evening, start back for home through Witte Poort, and so along the fairly level road that led along by the Plessis River up to our farmstead—a matter of eight or nine miles. Descending the mountain with some difficulty, for it was steep, broken, and crumbling, we reached the plain, and presently arrived at the store.

Squatted outside, blinking in the hot glare of the sunlight, were three or four Kaffirs, one of whom I recognised as a petty chief from a neighbouring kraal, making up their minds for a drink, or devising means of raising the wind. Inside, as we entered, another was having a *soupje*, a small tumbler of raw spirit—Boer brandy—which he gulped down without even winking. Not a pleasant sight, but, alas, a too frequent one at South African "winkels" (stores). Another fiery dram followed, gulped down, like its predecessor, quite unblenchingly, and yet that crude and awful spirit is enough to make even a dead donkey sneeze. Truly these natives must have "interiors" of cast iron. Now I am not straight-laced, or a preacher, or a practiser of total abstinence, but when I remember that the Hottentots, who 250 years ago held sway in this vast territory, are now, as a race, absolutely demoralised by strong drink; that out of their numbers probably more than fifty per cent. are at the present time in a greater or less degree drunkards; when I reflect that drink forms the unhappy beginning and the miserable end of the story of their intercourse with the white man; when, too, I see a magnificent race like the Kaffirs falling before the same evil influences, I feel sick at heart, and heartily wish every barrel of the filthy poison sold to these poor people a hundred fathoms deep beneath the ocean.

Having entered the "winkel," Jackson forthwith flung himself down on the top of an empty cask, and having ordered him some bottled beer and some sardines and biscuits, I followed the store-keeper into his inner sanctum, there to refresh myself and look at some Kaffir curios. Having finished a bottle of Bass, I came out into the store again, and there found Jackson discussing a third glass of "square-face" (Hollands), in succession to the beer I had paid for. This would not do at all, for I knew his head was anything but strong, and I had to get him home. Alas! the mischief was done, as I found to my cost later on. Jackson now lay down for a nap before we started, and I lighted a pipe, chatted to the store-keeper, and went out to have a look at his ostriches. In about three-quarters of an hour I roused Jackson, and prepared to set off homewards, for it was now four o'clock. I wanted to take a little treat home for Mrs. H., our hostess, and I therefore bought half a dozen bottles of stout, which, at two shillings and sixpence a bottle, is a luxury in these regions. Three of these I inserted into the capacious pockets of my own shooting jacket, the other three I got Jackson, who, though drowsy and tired, seemed pretty well, to carry in the same way.

Then we prepared to start. About three hundred yards from the house lay the drift or ford of the Groote River, and towards this we made our way. The drift was strewn with rough stones and boulders, and the water ran in a shallow stream between these; with ordinary care, it would not be a difficult matter to get across dryshod. I went over first, and when half-way, or a little more, turned to look for Jackson; alas! at that moment he stumbled and fell headlong into the stream, smashing my gun, which he carried, short off at the stock, and smashing also the three bottles of stout with which he was loaded up. I picked him up, put his hat on, took the pieces of the gun under my arm in addition to my own rifle, and helped the man across the remainder of the drift; he was simply dripping with stout, which leaked through his pockets and streamed down his legs. On the other bank, I picked the pieces of broken bottles from his pockets, and cleaned him down as well as possible; then, with his arm in mine, we continued the journey—a journey I shall never forget.

We proceeded about three hundred yards, and I was vexed to find that the fresh air had an entirely different effect upon Jackson from what I had hoped and expected. He became rapidly more helpless; he leaned heavily upon me, reeling in his gait, and occasionally letting his hat fall from his head. When

I picked it up he usually fell down; however, we must get home, and slowly and laboriously we proceeded. I suppose if Jackson dropped his hat and fell down once in the next two miles, he did it fifty times; it was really too disgusting. Presently I began to feel tired, for it is no joke, after a fifteen miles' mountain tramp, to have to carry a rifle, a gun broken in two, three bottles of stout, oneself, and a tipsy man into the bargain. With infinite labour we completed another mile in this manner, until, as dusk came on, I became exhausted. Here Jackson implored me to leave him in the bush to go to sleep. Another half-mile of falls and pickings up, and I sat down for a moment's rest beside the unfortunate Jackson, who instantly seized the opportunity to fall fast asleep. It was now dark; in the distance I could hear the noisy barking of the curs belonging to a Kaffir kraal.

I thought of taking Jackson there, but I scarcely liked the idea, for although the Kaffirs would look after him till morning, he would probably be placed in a hut swarming with vermin. Then I tried to wake him and proceed. In vain; he muttered, "Leave me to sleep," and I could do nothing with him; he would *not* wake. Finally I placed him under a dry thick bush away from the road, and left him to follow on in the morning when he had recovered; he could take no harm, and the nights were dry and not too cold. Wearily I resumed my journey, and about three miles from home, the last stroke of a day of evil luck overtook me.

I had carried all day, suspended to a belt, a sheathed hunting knife—one of Silver & Co.'s best specimens—for the purpose of giving the *coup de grace* to, and skinning game when wounded. The knife, a doubled-edged one, was nearly new, and its point and edges were as sharp and keen as they well could be. During the operations of stooping and picking up the unfortunate Jackson, the sheath had by some means come off, and as it was dark, to save time, I slipped it into my pocket, and very unwisely left the knife to hang naked at my side. Before long the belt had slipped round, and the knife hung, unknown to me, directly in my front. In crossing a dry spruit, and scrambling up the farther bank, I blundered in the dark, and making an extra effort, thrust my right thigh heavily against the point of the hunting knife. I knew in a moment what had happened ; the blade had gone pretty deeply into my leg, and directly I topped the bank I was bleeding freely, and felt the warm blood trickling down the wounded limb. Situated as I was, I could only bind my handkerchief tightly round my leg, trust to Providence, and hurry home. After a somewhat painful walk, I at length saw the welcome lights of the farmhouse, and presently got indoors ; then I found the leg of my right trouser soaked in blood.

Assistance was soon provided, the wound washed and tightly strapped up, and the bleeding presently ceased. An inspection showed that I had had a very narrow escape ; the wound was within one inch of the femoral artery; if this artery had been severed I should have bled to death very rapidly, and nothing could have saved me. An inch of a miss, however, in this case, was literally as good as a mile, and the only reminder I have of the adventure is a neat scar. I was about again in a few days. Jackson arrived at the farm at five o'clock in the morning, none the worse for his sleep in the veldt, but rather ashamed of himself. And so ended the most unlucky day's sport I ever experienced in Cape Colony.

CHAPTER XI.

A SECRET OF THE ORANGE RIVER.

MANY are the stories told at the outspan fires of the South African transport riders—some weird, some romantic, some of native wars, some of fierce encounters with the wild beasts of the land. Often, as I travelled with my friends up country, we stopped to have a chat with the rugged transport riders, and some strange and interesting information was obtained in this way. The transport rider—the carrier of Africa—with his stout waggon and span of oxen, travels, year after year, over the rough roads of Cape Colony, and beyond, in all directions, and is constantly encountering all sorts and conditions of men—white, black, and off-coloured; and in his wanderings, or over his evening camp-fire, he picks up great store of legend and adventure from the passing hunters, explorers, and traders.

One night, after a day's journey through the bush-veldt, we lay at a farmhouse, near which was a public outspan. At this outspan two transport riders were sitting snugly over their evening meal; they seemed a couple of cheery, good fellows—one an English Afrikander, the other an Englishman, an old University man, and well-read, as we afterwards discovered—and nothing would suit them but that we should join them, and take pot-luck. Attracted by their hospitable ways, and the enticing smell of their game stew, for we were none of us anthobians, we sat us down and ate and drank with vigorous appetites. Their camp-pot contained the best part of a tender steinbok, and a brace or two of pheasants (francolins); and we heartily enjoyed the meal, washed down with the inevitable coffee.

Supper finished, some good old Cango (the best home-manufactured brandy of the Cape, made in the Oudtshoorn district) was produced, pipes lighted, and then we began to "yarn." For an hour or more, we talked upon a variety of topics—old days in England, the voyage to the Cape, the Colony, its prospects and its sport. From these, our conversation wandered up country, and we soon found that our acquaintances were old interior traders, who in the days when ivory and feathers were more plentiful and more accessible than now, had over and over again made the journey to 'Mangwato and back. 'Mangwato, it may be explained, is the trader's abbreviation for Bamangwato, Khama's country, the most northerly of the Bechuana States; and of Bamangwato, Shoshong is the capital and seat of trade. Then we wandered in our talk to the Kalahari, that mysterious and little-known desert land, and from the Kalahari back to the Orange River again.

" 'Tis strange," said one of our number, "how little is known of the Orange River—at all events west of the falls; I don't think I ever met a man who had been down it. One would think the colonists would know something of their northern boundary; as a matter of fact they don't."

"Ah! talking of the Orange River, reminds me," said the younger of the transport riders, the ex-Oxonian, and the more loquacious of the two, " of a most extraordinary yarn I heard from a man I fell in with some years back, stranded in the 'thirst-land' north-west of Shoshong. Poor chap! he was in a sorry plight; he was an English gentleman, who for years had, from sheer love of sport and a wild life, been hunting big game in the interior. That season he had stayed too late on the Chobe River, near where it runs into the Zambesi, and with most of his people had got fever badly. They had had a disastrous trek out, losing most of their oxen and all their horses; and when I came across them they were stuck fast in the *doorst-land* (thirst-land) unable to move forward or back. For two-and-a-half days they had been without water; and from being in bad health to begin with, hadn't half a chance; and if I had not stumbled upon them, they must all have been dead within fifteen hours. I had luckily some water in my *vatjes*, and managed to pull them round; and that night, leaving their waggon in the desert, in hopes of being saved subsequently, and taking as much of the ivory and valuables as we could manage, and Mowbray's (the Englishman's) guns and ammunition, we made a good trek, and reached water on the afternoon of the next day. I never saw a man so grateful as Mowbray; I believe he would have done anything in the world for me after he had pulled round a bit. Poor chap, during the short time I knew him, I found him one of the best fellows and most delightful companions I ever met. Unlike most hunters, he had read much, and could talk well upon almost any subject; and his stories of life and adventure in the far interior interested and impressed me wonderfully; but the Zambesi fever had got too strong a hold upon him. I dosed him with quinine, and pulled him together till we got to Shoshong, where I wanted him to rest; but he seemed restless and anxious to get out into the open veldt again, and after a few days we started away. Before we had got half-way down to Griqualand, Mowbray grew suddenly worse, and died one evening in my waggon just at sunset. We buried him under a kameel doorn tree, covering the grave with heavy stones, and fencing it strongly with thorns to keep away the jackals and hyenas.

"Many and many a talk I had with poor Mowbray before he died; sometimes he would brighten up wonderfully, and insist on talking to me for hours, as he lay, well wrapped up, in the evening, underneath my waggon sail. One evening, in particular, he had seemed so much stronger and better—and in the evening, as we sat before the camp fire on the dewless ground, where I had propped him up and made him comfortable, he told me a most strange story, a story so wonderful that most people would scout and laugh at it as wildly improbable; yet, remembering well the narrator and the circumstances under which he told it to me, with the shadow of death creeping over the short remaining vista of his life, I believe most firmly his story to be true as gospel.

"Poor chap! he began in this way: "Felton, you have been a thundering kind friend to me, kind and tender as any woman (which, by the way, was all nonsense), and I feel I owe you more than I am ever likely to repay; yet, if you want wealth, I believe I can put it in your way. Do you know the northern bank of the Orange

River, between the great falls and the sea? No! I don't suppose you do, for very few people have ever trekked down it; still fewer have ever got down to the water from the great walls of desolate and precipitous mountain that environ its course, and except myself and two others, neither of whom can ever reveal its whereabouts, I believe no mortal soul upon this earth has ever set eyes upon the place I am going to tell you about. Listen!

" In 1871, about the time the diamond-fields were discovered, and people began to flock to Griqualand West, I was rather bitten with the mania, and for some months worked like a nigger on the fields; during that time I got to know a good deal about stones. I soon tired of the life, however, and finally sold my claim, and what diamonds I had acquired, fitted up a waggon, gathered together some native servants, and trekked again for those glorious hunting grounds of the interior, glad enough to resume my old and ever-charming life. Amongst my servants was a little Bushman, Klaas by name, whom I afterwards found a perfect treasure at spooring and hunting. Like all true Bushmen, he was dauntless as a wounded lion, and determined as a rhinoceros, which is saying a good deal. I suppose Klaas had had more varied experience of South African life than any native I ever met. Originally, he had come as a child from the borders of the Orange River, where he had been taken prisoner in a Boer foray, in which nearly all his relations were shot down. He had then been 'apprenticed' in the family of one of his captors, where he had acquired a certain knowledge of semi-civilised life. From the Boer family of the back country, he had subsequently drifted farther down into the Colony, and thence into an elephant hunter's retinue. He had accompanied expeditions with Griquas, Dutch, and Englishmen all over the far interior. The Kalahari desert, Ovampoland, Lake N'Gami, the Mababé veldt, and the Zambesi country, were all well known to him, for in all of them he had traded, hunted, and, on occasion, fought. As for the western Orange River and its mysteries—for it is a mysterious region—he knew it, as I afterwards discovered, better than any man in the world. Well, we trekked up to Matabeleland, and, after some trouble, got permission to hunt there; and a fine time we had, getting a quantity of ivory, and magnificent sport among lions, elephants, buffaloes, rhinoceros, sable and roan antelopes, koodoo, eland, Burchell's zebras, pallah, and all manner of smaller game.

"One day, Klaas, who was sometimes a bit too venturesome, got caught in the open by a black rhinoceros, a savage old bull. The old brute charged and slightly tossed him once, making a nasty gash in his thigh, but not fairly getting his horn under him, and was just turning to finish the poor little beggar, when I luckily nicked in. I had seen the business, and had had time t rush out on to the plain, and just as *Borélé* charged at poor Klaas to finish him off as he lay, I got up within forty yards, let drive, and, as luck would have it, dropped him with a ·500 express bullet behind the shoulder. Even then the fierce brute recovered himself, and tried to charge me in turn; but he was now disabled, and I soon settled his game. After that episode, Klaas proved himself about the only grateful native I ever heard of, and seemed as if he couldn't do enough for me.

"One day, after he had got over his wound, he came to me, and said, 'Sieur! you said one day that you would like to know whether there are diamonds anywhere else than at New Rush (as Kimberley was then called). Well, sieur, I have been working at New Rush, and I know what diamonds are like; and I can tell you where you can find as many of them in a week's search as you may like to pick up. Allemaghte! Ja, it is as true, sieur, as a wilde honde on a hartebeest's spoor.'

"'What the devil do you mean, Klaas!' said I, turning sharply round — for I was mending the dissel-boom (waggon pole)—to see if the Bushman was joking. But, on the contrary, Klaas's little weazened monkey-face wore an expression perfectly serious and apparently truthful. The statement seemed strange, for I knew the little beggar was not given to 'blowing,' as so many of the Kaffirs and Totties are.

"'Ja, sieur, it is truth; if ye will so trek with me to the Groote (Orange) Rivier, three or four days beyond the falls, I will show you a place where there are hundreds and hundreds of diamonds, big ones, too, many of them, to be found lying about in the gravel. I have played with them, and with other "mooi steins," too, often and often as a boy, when I used to poke about here and there, up and down the Groote Rivier. My father and grandfather lived near the place I speak of, and I know the way to the "vallei," where these diamonds are, well, though no one but myself knows of them; for I found them by a chance, and, selfish like, never told of my child's secret. I will take you to the place if you like.'

"'Are you really speaking truth, Klaas?' said I severely.

"'Ja! Ja! sieur, I am, I am,' he earnestly and vehemently reiterated, 'you saved my life from the "rhenoster" the other day, and I don't forget it.'

"Again and again I questioned and cross-questioned the little Bushman, and finally convinced myself of his truth, and I had too much respect for his keen intelligence to think he was himself misled or mistaken.

"'Well, Klaas,' said I at last, 'I believe you, and we'll trek down to the Orange River and see this wonderful diamond valley of yours.'

"Shortly after this conversation we came back to Shoshong, where I sold my ivory, and then with empty waggon, and the oxen refreshed by a good rest, set our faces for the river. From Shoshong, in Bamangwato, we trekked straight away across the south-eastern corner of the Kalahari, in an oblique direction, pointing south-west; it was a frightfully waterless and tedious journey, especially after passing the Langeberg, which we kept on our left hand. Towards the end of the journey, we found no water at a fountain where we had expected to obtain it, and thereby lost four out of twenty-two oxen (for I had six spare ones), and at last, after trekking over a burning and most broken country, we were beyond measure thankful to strike the river some way below the great falls. Klaas had led us to a most beautiful spot, where the terrain slopes gradually to the river (the only place for perhaps thirty or forty miles, where the water, shut in by mighty mountain walls, can be approached), and where we could rest and refresh ourselves and our oxen. Here we stopped four days. It was a lovely spot; down the banks of the river, and following its course, grew charming avenues of willows, kameel doorns (mimosa), and bastard ebony; two or three islands densely clothed with bush and greenery dotted the broad and shining bosom of the mighty stream; hippopotami wallowed quietly in the flood, and fish were plentiful. The mimosa was now in full bloom, and the sweet fragrance of its yellow flowers everywhere perfumed the air as one strolled by the river's brim. Rare cranes, flamingoes, gorgeous kingfishers, and many handsome geese, ducks, and other water-fowl, lent life and charm to this sweet and favoured oasis.

"I had some old scraps of fishing tackle with me, and having cut myself a rod from a willow tree, I employed some of my spare time in catching fish, and had, for South Africa—which, as you know, is not a great angling country—capital sport. The fish I captured were a kind of flat-headed barbel, fellows with dark greenish-olive backs and white bellies, and I caught them with scraps of meat, bees, grasshoppers, anything I could get hold of, as fast as I could pull them out, for an hour or two at a time. Once I ran clean out of bait, and was non-plussed; however, I turned over a stone or two, killed a couple of scorpions, carefully cut off their stings, and used them as baits, and the fish came at them absolutely like tigers. I soon caught some thirty pounds' weight of fish whenever I went out. The mountains rose here and there around in magnificently serrated peaks, and the whole place, whichever way you looked, was superbly beautiful. There was a fair quantity of game about; Klaas shot some klipspringer antelopes—hereabouts comparatively tame—up in the mountains, and there were koodoos, steinboks, and duykers in the bushes and kopjes.

"After the parching and most harassing trek across the desert, our encampment seemed a terrestial paradise. The guinea-fowls called constantly with pleasant metallic voices from the trees that margined the river, and furnished capital banquets when required. Many fine francolins abounded, and at evening, Namaqua partridges came to the water to drink in literally astounding numbers. We had to form a strong fence of thorns around us, for leopards were numerous and very daring, and there were still lions about in that country. At night, as I lay in my waggon, contentedly looking into the starry blue, studded with a million points of fire, and mildly admiring the glorious effulgence of the greater constellations, I began to conjure up all sorts of dreams of the future, of which the bases and foundations were piles of diamonds, culled from Klaas's wondrous valley.

"Having recruited from the desert journey, and all, men and beasts, being in good heart and fettle, we presently started away down the river for the valley of diamonds. I had, besides Klaas, four other men as drivers, *voer-lopers* and after-riders, and they naturally enough were extremely curious to know what on earth the 'Baas' could want to trek down the Orange River for—a country where no one came, and of which no one had ever even heard. I had to tell them that I was prospecting for a copper mine, for, as you probably know, there are many places in this region where that metal occurs. After our four days' rest by the noble river we were all greatly refreshed, and quite prepared for the severe travel that lay again before us. As we were doubtful whether we should find water at the next fountain that Klaas knew of, owing to the prevalence of drought—and as it was an utter impossibility (so Klaas informed me) to get down to the river on this side for several days owing to the steep mountain wall that everywhere encompassed it—I filled the water *vatjes*, and every other utensil I could think of; and then, all being ready, and the oxen inspanned, we moved briskly forward.

"We had now to make a *détour* to the right, away from the river, and for great part of a day picked our painful footsteps over a rough and semi-mountainous country. Towards evening, we emerged upon a dreary and interminable waste that lay outstretched before us, its far horizon barred in the dim distance by towering mountains, through which we should presently have to force our passage. That evening we outspanned in a howling wilderness of loose and scorching sand, upon which scarcely a bush or shrub found subsistence. After a night, not too comfortable, and broken by some hyenas that prowled restlessly about, we were up betimes next morning. As soon as the oxen were inspanned, and ready to move forward for the mountains to which Klaas had directed our course, I rode off for a low kopje that rose from the plain away in the distance, hoping to see game beyond. I was not disappointed; a small troop of hartebeest was grazing about half-a-mile off, and by dint of a little manœuvring with my

Hottentot after-rider, whom I despatched on a *détour*, I managed to cut across the herd, and knocked over a fat cow at forty yards. We soon had her skinned, and taking the best of the meat, rode on for the waggon. Again we had an exhausting trek over a burning sandy plain; the heat of this day was something terrible. I have had some baddish journeys in the *doorstland* on the way to the great lake, but this was, if possible, worse. Towards four o'clock, the oxen were ready to sink in their yokes; their lowing was most distressing, and as the water was now nearly at an end, and we might not reach a permanent supply for another day, nothing could be done to alleviate their sufferings. At nightfall, more dead than alive, we outspanned beneath the loom of a gigantic mountain range, whose recesses we were to pierce on the following morning. Half a day beyond this barrier lay the valley of diamonds, as Klaas whispered to me after supper that night with gleaming excited eyes; for, noticing my growing keenness, he, too, was becoming imbued with something of my expectancy.

"That night, as we lay under the mountain, was one of the most stifling I ever endured in South Africa, where, on the high table-lands of the interior, nights are usually cool and refreshing. Even the moist heat of the Zambesi Valley was not more trying than this torrid, empty desert. The oven-like heat, cast up all day from the sandy plain, seemed to be returned at night by these sun-scorched rocks with redoubled intensity. Waterless, we lay sweltering in our misery, with blackened tongues and parched and cracking lips. The oxen seemed almost like dead things. Often have I inwardly thanked Pringle, the poet of South Africa, for his sweet and touching verse, written with the love of this strange wild land deep in him, for his striking descriptions of its beauties and its fauna. As I lay panting that night, cursing my luck and the folly that brought me thither, I lit a lantern, and opened his glowing pages. What were almost the first lines to greet my gaze? These!

> 'A region of emptiness, howling and drear,
> Which man hath abandoned, from famine and fear,
> Which the snake and the lizard inhabit alone,
> With the twilight bat from the yawning stone;
> Where grass, nor herb, nor shrub, takes root,
> Save poisonous thorns, that pierce the foot;
> And here, while the night-winds around me sigh,
> And the stars burn bright in the midnight sky,
> As I sit apart, by the desert stone,
> Like Elijah at Horeb's cave alone,
> " A still, small voice " comes through the wild,
> (Like a father consoling his fretful child),
> Which banishes bitterness, wrath, and fear,
> Saying,—Man is distant, but God is near.'

"True! True! And so, thanks to Pringle, I bore up through that most miserable night, with a stouter heart, an easier mind."

Here the transport-rider paused in his story, got up, and diving into his waggon, pulled out a small volume, roughly bound in the soft skin of an Ourebi fawn. Opening its well-thumbed pages, he showed us the passage he had just quoted, deeply scored in pencil. "There," said he, "is the very book poor Mowbray read that night; there is the passage, marked with his own hand. Every line of it I know by heart, and his book I treasure greatly, and nearly always carry with me." Then he resumed his narrative.

"We hailed," said Mowbray, "the passage of the mountains next morning with something akin to delight; anything to banish the monotony of these last two days of burning toil. We were up as the morning star flashed above the earth-line. We drank the remaining water, which afforded barely half-a-pint each to the men, none for the oxen and horses. With difficulty the poor oxen, already, in this short space, gaunt and enfeebled from the heat, and for lack of food and drink, were forced up into their yokes. Klaas, as the only one of us who knew the country, directed our movements, and with hoarse shouts, and re-echoing cracks from the mighty waggon-whip, slowly our caravan was set in motion. Our entrance to the mountains was effected through a narrow and extremely difficult poort (pass), strewn with huge boulders, and overgrown with brush and underwood that often barred the way, and rendered stoppages frequent. After about a mile, the kloof into which this poort debouched suddenly narrowed and turned left-handed at right angles to our course. Accompanied by Klaas, I walked down it, and was soon convinced by the little Bushman that our passage that way was ended. As Klaas had warned me, our only way through and out of the mountains now lay in taking, with our waggon, to the steep and broken hillsides, a proceeding not only perilous, but apparently all but impossible. Yet the thing had to be done, and we at once set the spent oxen in motion, and faced the ascent obliquely. After consultation with Klaas, I got out some ropes which I had fastened to the uppermost side of the waggon, while some stout long poles which I had had previously cut for such an emergency, while outspanned at the Orange River, served to prop up our lumbering vehicle from the lower side. Slowly and wearily, and yet, withal, with a sort of dogged stubbornness, the poor oxen toiled on, half-hour after half-hour, urged by our shouts, by the cruel waggon-whip, mercilessly plied, and the terrible after-ox *sjambok*.* Many times it seemed, as our cumbrous desert ship crashed across a boulder, or down a stair-like terrace of rock, that it must inevitably topple over, and roll crashing to the bottom; but our guy-ropes, and the supporting poles, saved us again and again.

"I had fastened one of the ropes with a stout band of leather round the chest of my hunting horse, the other two ropes were held by the strongest of my servants and myself, while two other men held the poles against the lower side of the waggon as they stood down hill below it. My old horse, guided by a Bechuana boy, as usual, proved himself as sensible as any Christian, knew exactly what he had to do; and when we came to crucial points, and the waggon shivered as it were upon empty space, he and my Kaffir and I tugged away, while the fellows below shoved with might and main. And so time after time we averted a catastrophe, so dire that I shuddered to think of it; for in some places, if the waggon had gone, the wreck must have been irreparable, and the yoked oxen hurled with it in a broken and mangled heap to the bottom far below us. Well, occasionally halting for a blow, long hours of the most distressing labour I ever

---

* *Sjambok*, a straight, tapering, cutting whip, of rhinoceros or hippopotamus hide—terribly punishing.

experienced were at last got through; we had surmounted and left behind the first huge mountain-side, had plunged into a valley, had passed obliquely over the shoulder of another great mountain, and now halted in a deep and hollow kloof lying below a singular flat-topped mountain, conical in shape, that stretched across our onward path. This mountain was flanked on either hand, as we fronted it, by yawning cliffs, and was only approachable from this one aspect. Here we outspanned for a final rest before completing our work, if to complete it were possible. Shading my eyes from the fierce sunlight, I looked upward at the long slope of mountain, broken here and there, and occasionally shaggy with bush; over all, the fierce atmosphere quivered, seething and dancing in the sunblaze. I looked again with doubt and dismay at the gasping oxen, many of them lying foundered, and almost dead from thirst and fatigue, and my spirits, usually brisk and unflagging, sank below zero. Klaas had told me, previously, of a most wonderful pool of water that lay on the crown of a mountain, where we should outspan finally before entering upon the portals of the diamond valley. Now he came to me and said, pointing upwards, 'Sieur, de sweet water lies yonder *op de berg*. It is a beautiful pool, such as ye never saw the like of; if we reach it we are saved, and the oxen will soon get round again; ye must get them up somehow, even without the waggon.'

"The tiny yellow blear-eyed Bushman, standing over me as I sat on a rock, pointing with his lean arm skywards, his anxious dirt-grimed face streaming with perspiration, was hardly the figure of an angel of hope, and yet at that moment he was an angel, of the earth, earthy, 'tis true, yet an angel that held before us sure hope of rescue from our valley of despair; for despair, black and utter now, lay upon the faces of my followers, and in the eyes of my oxen. Remember, we had tasted no water to speak of for close on three days, and had had, besides, a frightfully trying trek.

"We lay panting and grilling for an hour or more, and then I told my men that water in any quantity lay at the mountain top, and that we must, at all hazards, get the oxen up to it. By dint of severe thrashing with the after-ox sjambok, we at last got the oxen on to their legs—all but two, which could not be made to rise—and then, leaving the waggon, but taking three or four buckets, we moved upwards. Only a mile of ascent, or a little more, lay before us; but so feeble were the oxen, that we had the greatest difficulty to drive them to the top, even without the encumbering waggon. At last we reached the krantz, and after a hundred yards walk upon its flat top, we came almost suddenly upon a most wonderful, and to us, most soul-thrilling sight. Dense bush of mimosa, thorn, spekboom, euphorbia, Hottentot cherry, and other shrubs grew around, here and there relieved by wide patches of open space. The oxen, getting the breeze and scenting water, suddenly began to display a most extraordinary freshness; up went their heads, their dull eyes brightened, and they trotted forward to where the brush apparently grew thickest.

"For a time they found no opening, but after following the circling wall of bush, at length a broad avenue was disclosed—an avenue doubtless worn smooth by the passage of elephants, rhinoceroses, and other mighty game, in past ages; and then there fell upon our sight the most refreshing prospect that man ever gazed upon. Thirty yards down the opening, there lay a great pool of water, about 200 feet across at its narrowest point, and apparently of immense depth; the pool was circular, its sides were of rock and quartz, and completely inaccessible from every approach save that by which we had reached it. It was indeed completely encompassed by precipitous walls about thirty feet in height, which defied the advent of any other living thing than a lizard or a rock-rabbit. Upon these rocky walls grew lichens of various colours—blood-red, yellow, and purple, imparting a most wonderful beauty to the place. The avenue to the brink of this delicious water was of smooth rock somewhat sloping, and in the rush to drink we had the greatest difficulty in preventing the half-mad oxen from plunging or being pushed in, in which case we should have had much trouble to rescue them.

"How the poor beasts drank of that cool pellucid flood, and how we human beings drank too; I thought we should never have finished. The oxen drank and drank till the water literally ran out of their mouths as they at last turned away. Then I cast off my clothes and plunged into the water; it was icy cold and most invigorating, and I swam and splashed to my heart's content. After my swim and a rest, I directed my men to fill the four buckets we had brought, and then, leaving the horses in charge of one of their number, we drove the cattle, loth though they were to leave the water, back to the waggon, going very carefully so as not to spill the water. At length we reached the valley, only to find our two poor foundered bullocks lying nearly dead. The distant lowing of their refreshed comrades had, I think, warned them of good news, and the very smell of the water revived them, and after two buckets apiece of the cold draught had been gulped down their kiln-dried throats, they got up, shook themselves, and rejoined their fellows.

"We rested for a short time, and then inspanned and started for the upland pool. The oxen, worn and enfeebled though they were, had such a heart put into them by their drink, and seemed so well to know that their watery salvation lay up there, only a short mile distant, that they one and all bent gallantly to the yokes, and dragged their heavy burden to the margin of the bush-girt water. We now outspanned for the night, made strong fires, for the spoor of leopards was abundant, stewed some bustards, ate a good supper, and turned in; when I say turned in, I should be more correct in saying *I* turned into my waggon, and the men wrapped themselves in their blankets or karosses, lay with their toes almost into the fire, and snored in the most varied and inharmonious chorus that ears ever listened to.

"I suppose we had not been asleep two hours, when I was awakened by the sharp barks and yelpings of my dogs, the kicks and scrambles of the oxen, and the shouts of the men. Snatching up my rifle and rushing out, I was just in time to see a firebrand hurled at some dark object that sped between the fires.

"'What is it, Klaas?' I shouted.

"'Allemaghte! it is a tiger* (leopard), sieur,' cried the Bushman, 'and he has clawed one of the dogs.'

"True enough, on inspecting the yelping sufferer, Rooi-kat, a brindled red dog, and one of the best of my pack, I found the poor wretch at its last gasp, with its throat and neck almost torn to ribbons. Nothing could save the unfortunate animal, the blood streamed from its open throat, and, after a convulsive kick or two, it stretched itself out, and lay there dead. Cursing the sneaking, cowardly leopard, I saw that the replenished fires blazed up, and again turned in.

"It must have been about two o'clock in the morning, the coldest, the most silent, and the dreariest of the dark hours—that fatal hour betwixt night and day, when many a flickering life, unloosed by death, slips from its moorings—when I was again startled from slumber by a most blood-curdling yell. Hunters, as you know, sleep light, and seem instinctively to be aware of what passes around them, even although apparently wrapped in profoundest sleep. I knew in a moment that that agonised cry came from a human throat, and headlong from my kartel† I dashed. God! what a din was there again from dogs, men, and oxen, and, above all, those horrid human screams. I had my loaded rifle, and rushing up to a confused crowd struggling near the firelight, I saw in a moment what had happened.

"The youngest of my servants, a mere Bechuana

---

* At the Cape the leopard is invariably known by the Dutch misnomer of "tiger."

† Kartel, a sleeping framework fitted up in Cape waggons.

boy, was hard and fast in the grip of an immense leopard, which was tearing with its cruel teeth at his throat, and at the same time kicking murderously with its heavily clawed hind legs at the poor fellow's stomach and thighs. One of the men—Klaas of course—bolder than his fellows, was lunging an assegai into the brute's ribs, seemingly without the smallest effect; others were thrashing at it with firebrands; and the dogs were vainly worrying at its head and flanks. All this I saw instantaneously. Thrusting my followers aside, I ran up to the leopard, and, putting my rifle to its ear, fired. The express bullet did its work at once; the fiercest and most tenacious of the feline race could not refuse to yield its life with its head almost blown to atoms, and loosening its murderous hold, the brute fell dead. But too late! the poor Bechuana boy lay upon the sand wounded to the death. His right shoulder and throat were terribly ribanded and mangled by the foreclaws and teeth of the deadly cat; but the cruellest wounds lay lower down. The hinder claws of the leopard had absolutely torn the abdomen away; it was a shocking sight. Recovery was hopeless, and, indeed, although we did what we could for the poor sufferer, he only lingered an hour insensible, then died. After his death my men told me how the thing had happened. In this solitary region, the leopards and other *feræ*, as I have often heard, never being disturbed by gunners, are extraordinarily fierce and audacious. The leopard—a male—was evidently very hungry, as its empty stomach testified, and after once tasting blood—that of the dog—it soon got over its temporary scare. The young Bechuana lay farthest from the fire, for his elders took up the warmest positions, and the leopard had crept cat-like in upon him, and got him by the throat before he knew where he was. Then came the awful shriek I had heard, and then began the tussle for life, alas! an altogether one-sided one. My men in the scramble—and scared too, no doubt—forgot the guns which were in the waggon, and only Klaas had thought of his assegai. So blood-thirsty was the brute, that nothing, except my rifle, could make it relax its hold, even although it was manifestly unable to get away with its victim. After these horrors, sleep was banished, and as the grey light came up, we prepared for day.

"The morning broke at length in ruddiest splendour, and as the terrain was slowly unfolded before my gaze, I realised the desolate magnificence of the country. Mountains, mountains, mountains, of grim sublimity rolled everywhere around. Far away below, as I looked westward, a thin silvery line, only visible for a little space, told of the great river flowing to the sea, inexorably shut in by precipitous mountain walls that guaranteed for ever its awful solitude.

"Klaas stood near, and as I gazed, he whispered, for my men were not far away: 'Sieur, yonder, straight in front of you, five miles away, lie the diamonds. If we start directly after breakfast, we shall have four hours hard climbing and walking to reach the valley.' 'All right, Klaas,' said I, 'breakfast is nearly ready, and we'll start as soon as we have fed.' A good fire was going, the pot was already steaming, the oxen had been watered, and I myself, stripping off my clothes on the brink of that delicious pool, dived deeply into its unknown depths. After a magnificent swim in the cold and bracing water, I felt transformed and ready for breakfast; but although the bathe had to some extent revived my spirits, I could not forget the sad beginning of our search—the death of poor Amazi, now, poor fellow, lying buried beneath a cairn of stones, just away beyond the camp.

"Well, breakfast was soon over, and then I spoke to my men. I told them that I intended to stay at this pool for a few days, and that in the meantime I was going prospecting in the mountains bordering the river. I despatched two of them to go and hunt for mountain buck in the direction we had come from, where we had noticed plenty of rhebok, duyker, and klipspringer; the others were to see that the oxen fed round about the water, where pasture was good and plentiful, and generally to look after the camp. For Klaas and myself, we should be away till dusk, perhaps even all night; but we did not wish to be followed or disturbed, and unless those at the camp heard my signal of four consecutive rifle shots, they were on no account to attempt to follow up our spoor. My men by this time knew me and my ways well, and I was convinced that we should not be followed by prying eyes; indeed, the lazy Africans were only too glad of an easy day in camp after their hard journey.

"Taking some biltong (dried flesh), biscuits, and bottle of water each, and each shouldering a rifle, Klaas and I started away at seven o'clock. The little beggar, who, I suppose, in his Bushman youth had wandered baboon-like all over this wild country till he knew it by heart, showed no sign of hesitation, but walked rapidly down hill into a deep gorge at the foot, which led half a mile or so into a huge mass of mountain that formed the north wall of the Orange River. This kloof must at some time or another have served as a conduit for mighty floods of water, for its bottom was everywhere strewn with boulders of titanic size and shape, torn from the cliff walls above. It took us a long hour of the most laborious effort to surmount these impediments, and then with torn hands and aching legs we went straight up a mountain, whose roof-like sides consisted of masses of loose shale and shingle, over which we slipped and floundered slowly and with difficulty. I say we, but I am bound to admit that the Bushman made much lighter of his task than I, his ape-like form seeming indeed much more fitted for such a slippery breakneck pastime.

"At length we reached the crest, and then, after passing through a fringe of bush and scrub, we scrambled down the thither descent, a descent of no little danger. The slipping shales that gave way at every step, often threatened indeed to hurl us headlong to the bottom, which we should most certainly have reached mere pulpy masses of humanity. At last this stage was ended, and we found ourselves in a very valley of desolation. Now we were almost completely entombed by narrowing mountain walls, whose dark red sides frowned upon us everywhere in horrid and overpowering silence. The sun was up, and the heat, shut in as we were, overpowering. Moreover, to make things more lively, I noticed that snakes were hereabouts more than ordinarily plentiful; the bloated puff-adder, the yellow cobra, and the dangerous little night adder, several times only just getting out of our path.

"The awful silence of this sepulchral place was presently, as we rested for ten minutes, broken by a posse of baboons, who having espied us from their krantzes above, came shoggling down to see what we were. They were huge brutes and savage, and quah-quahed at us threateningly, till Klaas sent a bullet into them, when they retreated pell-mell. We soon started again, and pressed rapidly along a narrow gorge some fifty feet wide with perfectly level precipitous walls, apparently worn smooth at their bases by the action of terrific torrents, probably an early development of the Orange River when first it made its way through these grim defiles. The ground we walked upon was, I noticed, composed of sand and rounded pebbles, evidently water-worn and of various kinds. Some of them were round masses of the most beautiful transparent crystal-spar, often as large as a man's head.

"Presently the causeway narrowed still more, and then turning a sharp corner, we suddenly came upon a pair of leopards sauntering coolly towards us. I didn't like the look of things at all, for a leopard at the best of times is an ugly customer, even when he knows and dreads firearms, and here probably the animals had never even heard the report of a gun.

"The brutes showed no intention of bolting, but stood with their backs up, their tails waving ominously, and their gleaming teeth bared in fierce defiance. There was nothing for it, either we or they must retreat, and having come all this frightful trek for the diamonds, I felt in no mood to back down, even to *Felis pardus* in his very nastiest mood. Looking to our rifles, we moved very quietly forward, until within thirty-five yards of the grim cats. They were male and female, and two as magnificent specimens of their kind as sun ever shone upon. The male had now crouched flat for his charge, and not an instant was to be lost; the female stood apparently irresolute. Noticing this, and not having time to speak, we both let drive at the charging male; both shots struck, but neither stopped him. The lady, hearing the report, and apparently not liking the look of affairs, incontinently fled. With a hoarse throaty grunt, the male leopard flew across the sand, coming straight at me, and then launched himself into the air. I fired too hurriedly my second barrel, and, for a wonder, clean missed, for in those days I seldom failed in stopping dangerous game; but these beggars are like lightning once they are charging. In a moment the yellow form was flying through space, straight at my head; I sprang to one side, and Klaas, firing again, sent the leopard struggling to earth, battling frantically for life amid sand and shingle, with a broken back. Lucky was the shot, and bravely fired, or I had probably been as good as a dead man ere this. Another cartridge soon finished off the fierce brute. We noticed, on inspection, that one of our first two bullets had ploughed up the leopard's nose, and glanced off the forehead; the other had entered the chest, and passed almost from end to end of the body, while the third had broken the spine. Klaas soon whipped the skin off the dead leopard, and hid it under some stones, and we then proceeded, the whole affair having occupied but twenty minutes.

"Another mile of this canal-like kloof brought us

to an opening, and here a most singular sight fell upon my vision. Hitherto we had been so shut in that the sun failed to penetrate between the narrowing cliffs—except, probably, for a short while as it passed immediately above them.

"Suddenly, as the gorge widened on either hand, a blaze of sunlight glowed and glistened on the upright walls to the left hand of us. As I looked thither, one of the most marvellous sights in nature was, in an instant, laid bare; a sight that few mortals, even in æons upon æons of the past, have ever gazed upon, in these remote and most inaccessible regions of the Orange River. The wall of mountain on our left stood up straight before the hot sunlight, a dark reddish-brown mass of rock—I suppose some five hundred feet in height—and then sloped away more smoothly to its summit that overlooked the river, as I should judge, about a mile distant. As we came out into the sunshine, Klaas, pointing to the cliff, ejaculated, in quite an excited way, 'De paarl! de paarl! kek, sieur, kek!' (The pearl! the pearl! look, sir, look!). Looking upwards at the mass of rock, my eye was suddenly arrested by a gleaming mass that protruded from the dead wall of mountain. Half-dazzled, I shaded my eyes with my hand, and looked again. It was a most strange and beautiful thing that I beheld, a freak of nature the most curious that I had ever set eyes on. The glittering mass was a huge egg-shaped ball of quartz, of a semi-transparent milky hue, flashing and gleaming, in the radiant sunshine, with the glorious prismatic colours that flash from the unlucky opal. But yet more strange, above the 'paarl,' as Klaas quaintly called it, and overhanging it, was a kind of canopy of stalactite of the same brilliant opalescent colours. It was wonderful! Klaas here began to caper and dance in the most fantastic fashion, and then suddenly ceasing, he said, 'Now, sieur, I will soon show you the diamonds; they are there,' pointing to a dark corner of the glen, 'right through the rock.'

"'What made you call that shining stone up there "De paarl"?' said I, as I gazed in admiration at the beautiful ball of crystal.

"'Well, sieur, I was once with a wine Boer at the Paarl down in the Old Colony, and a man told me why they called the mountain there "De paarl"; and he told me, too, what the pretty gems were that I saw in the young vrouw's best ring when she wore it; and I then knew what a paarl was, and that it came from a fish that grows in the sea. And I remembered then the great shining stone that I found up here, when I was a boy, on the Groote Rivier, and I thought to myself, "Ah! Klaas, that was the finest paarl ye ever saw, that near where the pretty white stones lay." I mean the diamonds yonder, sieur.'

"At last, then, we were within grasp of the famous stones, concerning whose reality I had even to the last had secret misgivings. It was a startling thought. Just beyond there, somewhere through the rock walls, whose secret approach at present Klaas only knew, lay 'Sindbad's Valley.' Could it be true? Could I actually be within touch of riches unspeakable; riches, in comparison with which the wealth of Crœsus seemed but a beggar's hoard?

"I sat down on a rock, and lit a pipe, just to

think it over and settle my rather highly-strung nerves. The Paarl, as I could now see, was an unique formation of crystal-spar, singularly rounded upon its face. It, and the glorious canopy of hanging stalactite above it, must have been reft bare by some mighty convulsion that had anciently torn asunder these mountains, leaving the ravine in which we stood.

"As we drank from our water-bottles, and ate some of the dried flesh and biscuits we had brought with us, I noticed Klaas's keen little eyes wandering inquiringly round the base of the precipice in our front. He seemed puzzled; and as we finished our repast, and lit our pipes again, he said: 'The hole in the rock that leads from this kloof to the diamonds should be over there'—pointing before him; 'but I can't quite make out the spot, the bushes have altered and grown so since I was here as a boy, years and years ago.'

"We got up and walked straight for the point he had indicated, and reached the foot of the precipice. All along here, where the sand and soil had been swept in bygone floods or had formed from the slow disintegration of fallen rock from above, cactus, euphorbia, aloe and brush grew thickly, and in particular the curious *Euphorbia candelabrum*, with its many-branching arms, stood prominent. The Bushman hunted hither and thither in the prickly jungle, with the fierce rapidity of a tiger-cat after a running guinea-fowl; but, inasmuch as he was sometimes prevented from immediately approaching the rock-wall, he appeared unable to hit off the tunnel that led, as he had formerly told me, to the valley beyond. Suddenly, after he had again disappeared, he gave a low whistle, a signal to approach, to which I quickly responded. Quietly pushing my way towards him, I was astonished to see within a small clearing a thick and high thorn fence, outside of which Klaas stood. Inside this circular kraal was a low round hut, formed of boughs and branches strongly and closely interlaced. Klaas was standing watching intently the interior of the hut, which seemed to be barred at its tiny entrance by a pile of thorns lying close against it.

"What could it mean, this strange dwelling, inaccessible as it seemed to human life? Klaas soon found a weak spot in the kraal fence, and pulling down some thorns, we stepped inside and approached the hut. Here, too, Klaas pulled away the dry mimosa thorns from the entrance, and was at once confronted by a tiny bow and arrow, and behind that by a fierce little weazened face. Instantly, my Bushman poured forth a torrent of his own language, redundant beyond expression with those extraordinary clicks of which the Bushman tongue seems mainly to consist. Even as he spoke, the bow and arrow were lowered, the little head appeared through the entrance, and the tiniest, quaintest, most ancient figure of a man I had ever beheld stood before us. Ancient, did I say? ancient is hardly a meet description of his aspect. As he stood there, blinking like an owl in the fierce sunlight, his only covering a little skin kaross of the red rhebok fastened over his shoulders, standing not more than three feet eight or ten inches in height, he looked indeed coeval with the rocks around him. I never saw anything like it. Poor little oddity; dim though his eyes were waxing, feeble though his shrivelled arm, dulled though his formerly acute senses, he had, with all the desperate pluck of his race, been prepared to do battle for his hearth and home.

"In his own tongue, Klaas interrogated this antediluvian Bushman, and then, suddenly, as he was answered by the word ' 'Ariseep,' a light flashed across his countenance. Seizing his aged countryman by the shoulders, he turned him round and carefully examined his back. Lifting the skin kaross, and rubbing away the coating of grease and dirt that covered the right shoulder, Klaas pointed to two round white scars just below the blade-bone, several inches apart. Then he gave a leap into the air, seized the old fossil by the neck, and shrieked into his ears the most wonderful torrent of Bushman language I have ever heard. In his turn the old man started back, scanned Klaas intently from head to foot, and, in a thin pipe, jabbered at him almost as volubly.

"Finally, Klaas enlightened me as to this comical interlude. It seemed incredible; but this old man, 'Ariseep by name, was his grandfather, whom he had not set eyes on since, long years before, the Boer Commando had broken into his tribal fastness, slain his father and mother and other relatives, and carried himself off captive. The old man before us had somehow escaped in the fight, had crept away, and after years of solitary hiding in the mountains around, had somehow penetrated to this grim and desolate valley, where he had subsisted on Bushman fare—snakes, lizards, roots, gum, bulbs, fruit, and an occasional snared buck or rock-rabbit; these, and a little rill of water that gushed from the mountain-side hard by, supplied him with existence.

Here he had lingered for many years, alone and isolated. His only fear had been, as he grew older and feebler, the leopards infesting the neighbouring mountains. Against their attacks had he built the strong thorn fence, carefully closed at night, and the door of thorns, which he wedged tightly into the entrance-way.

"A strange meeting indeed it was, but after all not stranger than many things that happen in the busy world. So far as I could learn from Klaas, who himself was between forty and fifty, the ancient figure before us was laden with the burden of more than ninety years. Think of it! Ninety summers of parched Bushmanland, of burning Orange River mountains; ninety seasons of hunger and thirst and dire privation; great part of the earlier period varied by raids on the flocks of the Boers, and battles for existence with the wild beasts of the land!

"After nearly an hour's incessant chatter, during which I believe Klaas had laid before his monkey-like ancestor an epitomized history of his life, he told the old man we wished to get through the mountain, and that he had lost the tunnel of which he had known as a boy. 'Ariseep, who it seems, in the years he had been there, had explored every nook and cranny of the valley, knew at once what he meant, and quickly pointed out to us, not a hundred paces away, a dense and prickly mass of cactus and euphorbia bush; here, after half an hour's hewing and slashing with our hunting knives, we managed to open a pathway, and at last a cave-like opening in the mountain, about seven feet in diameter, lay before us. Grandfather

'Ariseep, questioned as to the tunnel, said, that upon first discovering it—which he had done quite by accident while hunting rock-rabbits—he had once been through, years before, and as he had found the passage long and dangerous, and the valley beyond appeared to him less interesting than his present abiding place, he had never repeated the journey. However, he gave us warning that snakes abounded, and might not impossibly be encountered in the twenty minutes' crawl, which, as Klaas had told me, it would take to get through. This opinion, translated by Klaas, was not of a nature to fortify me in the undertaking, yet, rather than leave the diamonds unexploited, I felt prepared to brave the terrors of this uncanny passage.

" It was now three o'clock; the sun was marching steadily across the brassy firmament, on his eastward trek, and we had no time to lose.

" 'In you go, Klaas,' said I, and, nothing loth, Klaas dived into the bowels of the mountain, I at his heels. For five minutes, by dint of stooping, and an occasional hands-and-knees creep upon the flooring of the tunnel, sometimes on smooth sand, sometimes over protruding rock and rough gravel, we got along very comfortably. Then the roof of the dark avenue—for it was pitch dark now— suddenly lowered, and we had to crawl along, especially I, as being taller and bulkier than Klaas, like serpents, upon our bellies. It was unpleasant, deuced unpleasant, I can tell you, boxed up like this beneath the heart of the mountain. The very thought seemed to make the oppression a million times more oppressive. It seemed that the frightful pile of rock, towering far above us, was bodily descending to crush us into a horrible and hidden tomb. The thought of lying here, squeezed down till Judgment Day, was appalling ; or, perhaps, more mercifully, one's bones might, ages hereafter, be discovered, as these regions became settled up, in much the same state in which mummified cats are occasionally found in old chimneys and hidden closets, when ancient dwellings are pulled down in England. Even Klaas, plucky Bushman though he was, didn't seem to relish the adventure, and spoke in a subdued and awe-stricken whisper. Sometimes since, as I have thought of that most gruesome passage, I have burst into a sweat, nearly as profuse, though not so painful, as I endured that day. At last, after what seemed to me hours upon hours of this painful crawling and Egyptian gloom, we met a breath of fresher air ; the tunnel widened and heightened, and in another five minutes we emerged into the blessed sunlight. Little Klaas looked pretty well 'baked,' even in his old leather 'crackers'* and flannel shirt ; as for myself, I was literally streaming ; every thread on me was as wet as if I had plunged into a river. We lay panting for awhile upon the scorching rocks, and then sat up and looked about us.

"If the Paarl Kloof, as Klaas called it, from whence we had just come, had been sufficiently striking, the mighty amphitheatre in which we lay was infinitely more amazing. Imagine a vast arena almost completely circular in shape, flat and smooth, and composed as to its flooring of intermingled sand and gravel, reddish-yellow in colour. This arena

---

* A Colonial term for leather trousers.

was surrounded by stupendous walls of the same ruddy brown rock we had noticed in Paarl Kloof, which here towered to a height of close on a thousand feet. An inspection of these cliffs, which sheered inwards from top to bottom, revealed the fact previously imparted to me by Klaas, that no living being could ever penetrate hither save by the tunnel passage through which we had come. The amphitheatre, which here and there bore upon its surface a thin and scattered covering of bush and undergrowth, seemed everywhere about half-a-mile across from wall to wall. In the centre of the red cliffs, blazing forth in splendour, ran a broad band of the most glorious opalescent rock-crystal, which flashed out its glorious rays of coloured light as if to meet the fiery kisses of the sun. This flaming girdle of crystal, more beautiful a thousand times than the most gorgeous opal, the sheen of a fresh-caught mackerel, or the most radiant mother-of-pearl, I can only compare in splendour to the flashing rainbows formed over the foaming falls of the Zambesi, which I have seen more than once. It ran horizontally and very evenly round at least two-thirds of the cliff-belt that encircled us. It was a wonderful and amazing spectacle, and I think quite the most singular of the many strange things (and they are not few) I have seen in the African interior.

"Well, we sat gazing at this crystal rainbow for many minutes, till I had somewhat feasted my enraptured gaze; then we got up, and at once began the search for diamonds. Directly I saw the gravel—especially where it had been cleansed in the shallow spruits and dongas by the action of rain and flood—I knew at once we should find 'stones';

it resembled almost exactly the gravel found in the Vaal River diggings, and was here and there strongly ferruginous, mingled with red sand, and occasionally lime. I noticed quickly that agates, jaspers, and chalcedony were distributed pretty thickly, and that occasionally the curious banddoom stone, so often found in the Vaal River with diamonds, and, indeed, often considered by diggers as a sure indicator of 'stones,' was to be met with. In many places the pebbles were washed perfectly clean, and lay thickly piled in hollow water-ways; here we speedily found a rich harvest of the precious gems. In a feverish search of an hour-and-a-half, Klaas and I picked up twenty-three fine stones, ranging in size from a small pigeon's egg to a third of the size of my little finger nail. They were all fine diamonds, some few, it is true, yellow or straw-coloured, others of purest water, as I afterwards learned, and we had no difficulty in finding them, although we wandered over not a twentieth part of the valley. I could see at once from this off-hand search that enormous wealth lay spread here upon the surface of the earth; beneath probably was contained fabulous wealth. I was puzzled at the time—and I have never had inclination or opportunity to solve the mystery since—to account for the presence of diamonds in such profusion. Whether they were swept into the valley by early floodings of the Orange River through some aperture that existed formerly, but had been closed by volcanic action, or whether, as I am inclined to think, the whole amphitheatre is a vast upheaval from subterraneous fires of a bygone period, is to this hour an unfathomed secret. I rather incline to the latter theory, and believe that like the Kimberley 'pipe,'—as diggers call it—the diamondiferous earth had been shot upwards funnel-wise from below, and that ages of floods and rain-washing had cleansed and left bare the gravel and stones I had seen upon the surface.

"From the search we had had, I made no doubt that a fortnight's careful hunting in this valley would make me a millionaire, or something very like it. At length I was satisfied, and as the westering sun was fast stooping to his couch, with a light heart and elastic step I turned with Klaas to depart. The excitement of the 'find' had quite banished the remembrance of that awful tunnel passage so recently encountered.

"'We'll go back now, Klaas,' said I, 'sleep in your grandfather's kraal, and get to the waggon first thing in the morning; then I shall arrange to return and camp a fortnight in Paarl Kloof, leaving the waggon at the pool. In that time we shall be able to pick up diamonds enough to enrich ourselves, and all belonging to us, for generations. I don't mind, then, who discovers the valley; they can make another Kimberley of it if they choose for aught I care.'

"At half-past five we again entered the tunnel. It was a nasty business when one thought of it again, but it would soon be over. As it flashed across my brain, I thought at the moment that two such journeys a day for six or seven days would be quite as much as even the greediest diamond lover could stomach. As before, Klaas went first, and for half the distance all went well. Suddenly, as we came to a sandy part of the tunnel, there was a scuffle in front, a fierce exclamation in Bushman language, and then Klaas called out in a hoarse voice: 'Allemaghte, sieur, een slang het mij gebissen' (Almighty, sir, a snake has bitten me). Heavens, what a situation! Cooped up in this frightful burrow, face to face with probably a deadly snake, which had already bitten my companion! Almost immediately Klaas's voice came back to me in a hoarse guttural whisper, 'I have him by the neck, sieur; it is a puff-adder, and his teeth are sticking into my shoulder. If you will creep up and lay hold of his tail, which is your side of me, we can settle him; but I can't get his teeth out without your help.' As you will remember, the puff-adder's striking fangs are very curved, and are often difficult to disengage once it has made its strike. Poor Klaas, I felt certain his days must be numbered, but there was nothing for it; I *must* help him.

"Crawling forwards, and feeling my way with fright-benumbed fingers, I touched Klaas's leg. Then, softly moving my left hand, I was suddenly smitten by a horrible writhing tail. I seized it with both hands, and finally gripped the horrid reptile (which I felt to be swollen with rage, as is the brute's habit) in an iron grasp with both hands. Then I felt, in the black darkness, Klaas take a fresh grip of the loathsome creature's neck, and, with an effort, disengage the deadly fangs from his shoulder. Immediately, I felt him draw his knife, and, after a struggle, sever the serpent's head from its body. The head he pushed away to the right, as far out of our course as possible, and then I dragged the writhing body from him, and, shuddering, cast it behind me as far as possible.

"At that moment I thought that, for the first

time in my life, I must have swooned. But, luckily, I bethought me of poor, faithful Klaas, sore stricken; and I called to him, in as cheerful a voice as I could muster, 'Get forward Klaas, for your life, as hard as you can, and, please God, we'll pull you through.'

"Never had I admired the Bushman's fierce courage more than now. Most men would have sunk upon the sand, and given up life and hope. Not so this aboriginal. 'Ja, sieur; I will loup,' was all he said.

"Then we scrambled onward, occasionally halting as the deadly sickness overtook Klaas; but all the while I pushed him forwards, and urged him with my voice. At last the light came, and, as my poor Bushman grew feebler and more slow, I found room to pass him, and so dragged him behind me to the opening into Paarl Kloof. Here I propped him for a moment on the sand outside, with his back to the mountain, and loudly called ''Ariseep' while I got breath for a moment.

"The sun was sinking in blood-red splendour behind the mountains, and the kloof and rock walls were literally aglow with the parting blush of day. Nature looked calm and serenely beautiful, and hushed in a splendour that ill-accorded with the agitating scene there at the mouth of the tunnel. All this flashed across me as I called for the old man. I looked anxiously at Klaas, and examined his wound; there were two deep punctures in the left shoulder, and from his having had to use some degree of force to drag off the reptile, the orifices were more torn than is usual in cases of snake-bite. Klaas was now breathing heavily, and getting dull and stupefied. I took him in my arms, and carried him to 'Ariseep's kraal, whence the old man was just emerging. At sight of his grandfather, Klaas rallied, and rapidly told him what had happened, and the old man at once plunged into his hut for something. Then Klaas's eyelids drooped, and he became drowsy—almost senseless. In vain I roused him, and tried to make him walk, and so stay the baleful effects of the poison, now running riot in his blood; he was too far gone. 'Ariseep now re-appeared with a small skin bag, out of which he took some dirty-looking powder. With an old knife he scored the skin and flesh around Klaas's wound, and then rubbed in the powder. I had no brandy or ammonia to administer, and therefore let the old Bushman pursue his remedy, though I felt, somehow, it would be useless. So it proved; either the antidote, with which I believe Bushmen often do effect wonderful cures, was stale and inefficacious, or the poison had obtained too strong a hold. My poor Klaas never became conscious again, though I fancied eagerly that he recognised me before he died, for his lips moved as he turned to me once. His pulse sank and sank, his face became dull and ashen, his eyelids quivered a little, his breath came hard and laboured, and at last, within an hour-and-a half from the time he was bitten, he lay dead.

"So perished my faithful and devoted henchman; the stoutest, truest, bravest soul that ever African sun shone upon. I cannot express to you the true and unutterable grief I felt, as with old 'Ariseep, I buried poor Klaas when the moon rose that night. We placed him gently in a deep sandy spruit, and over the sand piled heavy stones to keep the vermin from him. Then laying myself within 'Ariseep's kraal, I waited for the slothful dawn. As it came, I rose, called 'Ariseep from his hut, and bade farewell to him as best I could, for we neither of us understood one another. I noticed, by-the-bye, that no sign of grief seemed to trouble the old man. Probably, he was too aged, and had seen too much of death to think much about the matter.

"The rest of my story is soon finished. I made my way back to camp, told my men what had happened, and indeed took some of them back with me to Klaas's grave, and made them exhume his body to satisfy themselves of the cause of death; for these men are sometimes very suspicious; then we covered him again securely against wandering beasts and birds.

"I trekked back to the Old Colony, sold off my things, and came home. The diamonds I had brought away, realised in England £22,000. I have never dreamt of going to the fatal valley again; nothing on earth would tempt me, after that ill-starred journey, heavy with the fate of Klaas, and the Bechuana boy, Amazi. As for the tunnel, I would not venture once more into its recesses for all the diamonds in Africa, even if they lay piled in heaps at the other end of it. Except old 'Ariseep, Klaas had no relation that I knew of, and it was useless to think of spending the diamond money in that quarter. The old fellow had, so far as I could make him understand me, utterly refused to accompany me from the kloof, where he evidently meant to end his days; even if he had come, what could I have done for him? At his time of life, and with his peculiar habits, he could hardly have begun the world again, even if I had brought him home, bought him a country house, taken rooms in Piccadilly, dressed him in the height of fashion, and launched him upon society.

"Therefore, I left him as I found him. Klaas I have never ceased to mourn, from that day to this. Part of the £22,000 I invested for some relatives, the balance that I kept suffices, with what I already possessed, for all possible wants of my own. Then I came back to my dearly-loved South Africa for the last time; and a few years later, made the journey to the Chobe River, from which you rescued me in the thirst-land."

Such was the story related to us by the transport rider, in a clear and singularly graphic manner, to which these pages do scant justice. Our narrator wound up by telling us that Mowbray had further imparted to him the exact locality of the diamond valley, but, he added: "I have never yet been there, nor do I think that, for the present, it is likely I shall. Some day, before I leave the Cape, I may have a try and trek down the Orange River; but I don't feel very keen about that secret passage, after poor Mowbray's experiences."

* * * * *

We had sat wrapt listeners, for some hours of that soft, calm, African night. The glorious stars looked out from above us in their deep blue dome; the Southern cross shone in serene effulgence, as if too its sparkling gems claimed an interest in the legend of the lost diamonds. It was now two o'clock, and the camp fire of the transport riders burned low; just one more *soupje* we had with our friendly entertainers, and then, with hearty expressions of thanks and goodwill, rose to seek our beds. That night, before falling asleep, I pondered long upon the strange narrative we had heard. Often since I have done so. Often, too, have I thought of the lone grave of the English hunter, Mowbray, far out on dim Kalahari.

Many will think a grave in some green and umbrageous village churchyard in Old England a fitter resting place; I doubt it. I believe that for spirits, such as Mowbray; those hardy adventurers and bold hunters, who have opened up the South African interior, and have made the name of Englishman known and respected, even amongst the uttermost tribes, loving, as they ardently love, the wild free life of the desert, a last resting-place upon the open veldt would accord more with their habits and ideas. Thus, in the quiet Kalahari, amid the fauna, and the scenes he loved so well, Mowbray will sleep in peace. There are men—and if you meet them anywhere, you meet them in South Africa—who even in these latter days have in their blood the pure hunter spirit, possessed by rude and remote ancestors, who subsisted by their weapons, and lived only for the joy of war and of the chase. These modern men have, however, commingled with the old hunter spirit, and almost unknown to themselves, a love for the beautiful, for nature in her wildest, most primeval moods, and to these South Africa has offered a veritable paradise. These are the men who, of all others, really and truly enjoy existence, albeit in a rude kind of way. No one, who has not met South African hunters, British or Dutch, can rightly appreciate the overpowering sentiment that enthrals them. They, themselves, know not exactly what it is that binds them so passionately to a life not seldom of inordinate toil and privation. Once the veldt fever has bitten them, they can never entirely forget or forsake their semi-civilised existence. Perils, wounds, fevers, hunger, thirst, and hardships of every kind; none of these can quench the love they bear for their self-chosen hunter's life. The open air! The open veldt! Ever-changing scenery; everywhere a glorious game country; everywhere a spice of danger; and beneath it all, burning brightly in their ardent souls, lies, I am convinced, a keen — if to themselves a half-hidden, half-comprehended — love of the beautiful, oftentimes purer and more true than the pallid sentiment, the jargon of nonsense, heard but too often in the crowded picture galleries of cities.

CHAPTER XII.

THE FALL OF THE ELEPHANT.

WHILE travelling up country from Port Elizabeth, I often heard of the elephants that, thanks to a timely preservation, yet remain to Cape Colony. These animals wander in small troops through a large expanse of the dense bush-veldt that clothes much of the eastern portions of the Colony, and in the Knysna Forest to the extreme south; and many amusing stories are told of their unwieldy pranks upon the outskirts of civilisation. As we passed through Uitenhage, we heard that some of these obtrusive mammoths had been, a few days previously, destroying the telegraph posts, and over-running the railway line close to the town. Only last year (1888) a troop of sixty elephants—besides calves—was seen in the Addo Bush. Many of the farmers of the Winterhoek and other adjacent localities are visited by these playful giants, and occasionally awake in the morning to find their mealie fields and gardens severely punished. Two years since, in 1887, Mrs. Hess, of Vaaldam, Winterhoek, was awakened one night by the angry barking of her dog. Sallying forth she searched her premises, and finding everything quiet returned to bed; at two o'clock in the morning, however, a Hottentot servant aroused her with the curious intelligence that a wool-waggon was in the goat kraal. Naturally anxious to verify such perplexing tidings, Mrs. Hess at once went out to the kraal, and there beheld one of her finest goats flying in the air, and the rest of the flock rushing forth helter-skelter through a large breach in the enclosure. The startled lady saw also in the bright moonlight not a wool waggon, but a huge elephant carrying long gleaming tusks, and very sensibly she at once fled indoors. Next morning it was found that besides slaying the inoffensive goat, the elephant had trampled and destroyed the crops around the house, and had made havoc with a quantity of sheep dip. Upon the morning of the same day the District Inspector of Sheep Scab, while making his rounds, happened to come upon a troop of eight elephants in the same neighbourhood, but not deeming such mighty cattle within the purview of the Act he administers, he turned his horse's head and hurriedly rode away.

One of the last of the professional ivory-hunters who pursued their dangerous calling in this part of the Colony was one Thackwray, a most silent and expert tracker, a man of extraordinary courage and determination—amounting sometimes even to rash bravado.

Of this man, who was killed by an elephant somewhere about 1830, some extraordinary tales are related. I have been told, on excellent authority, that Thackwray met his death in the act of winning a most foolhardy bet. He wagered that he would pick out the biggest and most savage bull elephant of the adjacent forest, would chalk with his own hand a large cross upon the hind quarters of the animal, and would then shoot it and obtain its ivory. The first part of the bet was actually performed; the elephant was followed, chalked, and laid low, but, rising again before Thackwray had reloaded, it turned upon him, thrust one tusk through his thigh, then grasping him with its trunk, smashed the unfortunate man almost to atoms, and then falling upon him repeatedly with its knees, crushed his body to a pulp. The elephant was afterwards found dead with the chalk marks upon it, still uneffaced.

It is a remarkable thing that in modern times no attempt has been made to domesticate and utilise the African elephant, as with its Indian congener. Within this century, the young of the elephant might have been captured even in the Old Colony itself, but no Governor of the Cape has apparently ever considered the matter worth thinking of, or if he has, his thoughts have been dexterously concealed. In India the elephant, although inferior in stature and strength to the elephant of Africa, has been utilized to a very large extent, and its mighty thews and sinews and wonderful intellect have aided in many a scheme of useful work, and in many a distant campaign.

In South Africa its services could have been at least as advantageously employed; and yet decade after decade these magnificent creatures have been ruthlessly shot down and exterminated, until now it is almost too late to think of subjugating and employing them. But even now there is time, and within thirty miles of Port Elizabeth, a town of seventeen thousand inhabitants—the most thriving business emporium of eastern South Africa—a beginning might be made. I say *might*, for although the raw material is there, I doubt greatly if a commencement will now ever be made of this business; and yet, by the importation of a few Indian elephants and their trainers, and the establishment of proper capturing compounds, the whole thing might be successfully accomplished almost within sound of Port Elizabeth—certainly of Uitenhage.

This modern neglect of the African elephant is not a little singular, especially having in view the shining examples of our Indian dependency. The figures on many a Roman coin and medal bear ample testimony to the frequent user of *Elephas Africanus* in ancient times. This fact is quite indisputable, for not only do the huge flap ears—differing so entirely from the Indian species—extending to the point of the shoulder, instantly proclaim the African elephant's identity upon Roman coinage, but history gives at least two well-known instances of the employment of trained African elephants, even within Europe itself. Pyrrhus, king of Epirus, so far back as 280, B.C., aided by Egyptian Ptolemy, with men and elephants, successfully made war upon Rome, and chiefly by the help of his African elephants, obtained victory at Heraclea, upon the Gulf of Tarentum, and two years later, 279, B.C., at Asculum. In the second Punic War, Hannibal, starting from New Carthage, B.C., 218, made his never-to-be-forgotten passage of the Pyrennees and Alps, with thirty-seven African elephants; and these huge creatures, subsequently reinforced by more of their kind, in 213, B.C., undoubtedly contributed in no slight degree towards his memorable victories of the Trevia, Lake Thrasymene, Cannæ, Herdonea,

Centenius, upon the Anio, and in Lucania, in that wonderful sixteen years' Italian campaign.

In that long and most memorable war, the African elephant never saw defeat. Polybius tells us that no Roman General ever dared to close with Hannibal. May we not infer that the elephant contributed in great measure to this wholesome dread of the Carthaginian's prowess? Alas! how have the mighty fallen! Instead of assisting great captains and mighty armies, and contributing to splendid victories as of yore, the African elephant has been hunted and destroyed these two hundred years past with such unrelenting ardour, that his disappearance, not only from South Africa, but from the whole of the Dark Continent, is now only a question of a generation or two.

But a few short years back, I myself witnessed many a waggon-load of ivory passing down from the interior to the coast; nowadays the supply is meagre indeed, and the stream that erstwhile was poured forth in abundance, now trickles feebly and intermittently. It was a wonderful sight (albeit a melancholy, if one only reflected on the waste of life) to witness the great waggons, crammed with ivory, toil-worn and travel-stained by the storm and stress of thousands of miles of rough and uneasy trekking, standing outspanned for the evening, or discharging their rich contents in the market-place. What strange and moving histories those waggons could have yielded if but they could have found tongue —the silent witnesses of many a toilsome trek through arid thirst-land, dense jungle and park-like champaign, studded thickly with the giraffe-acacia, the baobab, and many another tree and shrub.

They could tell of many a moving battle with wild beasts, many a gallant head of game, and of the death of many a stout oxen and horse, and even perchance of the indefatigable hunters that made of them their home. But the glory and the passionate excitement of elephant hunting is rapidly departing from Southern Africa, never again to return.

Nothing can better illustrate the rapidity with which the wisest and most powerful of the brute creation is being wiped from the face of Africa, than a survey of his decline and fall within the broad territories contained between the Zambesi and the Indian and South Atlantic Oceans, during the sixty years last past.

At the beginning of this century the elephant was found plentifully upon the eastern confines, and in the southern forest-belt of the Cape Colony. As lately as 1830 ivory-hunters pursued their calling in the dense bush-veldt of the Eastern province, and still later in Kaffraria; but the Cape Government, foreseeing the probability of the mighty beast's extinction within a few years, at length proclaimed measures of protection; and it is a curious fact that there exist at the present moment, within the south-eastern limits of Cape Colony—within sight of the Indian Ocean—more wild elephants than are to be found for probably 1,500 miles inland. In the Addo Bush, not far from Port Elizabeth; in the Knysna and Zitzikamma Forests, and upon the jungly slopes of the Winterhoek Mountains, troops of elephants yet wander free and undisturbed, as also does the fierce and gloomy buffalo, and that king of antelopes, the koodoo. Beyond the Kei, in

Kaffraria itself, where fostering protective laws have not extended, the elephant is extinct.

Passing eastward through Natal, once abounding in, but now devoid of, elephants, we reach Zululand. In that territory between 1850 and 1875 immense numbers of the great pachyderm were hunted and destroyed. Baldwin, Drummond, and other well-known Nimrods, besides a numerous array of obscure but quite as deadly Dutch and colonial professional hunters, have practically wrought extinction in the Zulu country. North of the Zulus, in Amatongaland, the same state of things exists, with this difference: that some numbers of elephants, driven by the incursive gold-diggers and prospectors of Swaziland from their ancient secluded haunts in that territory, have recently trekked south across the Amatonga border. Here their extermination must soon follow.

It is probable that the country most abounding in the poor remnant of elephants now south of the Zambesi is the unhealthy region lying east and north-east of the Transvaal border—much of it known as Umzilaland. Here the prevalence of deadly fever in the hot months, and the tse-tse fly, have alone prevented Dutch hunters from completing their work of destruction; but even here the supply of elephant-life is now sparse and limited, and cannot long hold out. The Orange Free State, from the treeless open nature of its terrain, although formerly crowded with other game, was never a haunt of elephants. But not so with the Transvaal. Here, in 1837, Captain Cornwallis Harris, one of the first to explore the beautiful but unknown wilds then held by the fierce Moselikatse and his Matabeles, found elephants in astounding plenty. In one valley alone, he saw wandering, in peaceful seclusion, hundreds of the great mammals. But close on the heels of Harris followed the Boer Voer-trekkers, who, having attacked and driven out Moselikatse, turned their attention to the extraordinary wealth of ivory within their new-found borders. Their labours have been but too successful; and there now remains to the vast territories of the South African Republic probably not one solitary wild elephant! All have vanished. North of the Transvaal lie Matabele and Mashona lands, the country now ruled by Lobengula, son of the dreaded Moselikatse. So recently as between 1871 and 1875, these lands were tenanted by vast numbers of elephants; but, as Mr. F. C. Selous, the well-known hunter, tells us, it is now difficult in a year's hunting to come across a single elephant in these countries. Mr. Selous, himself the mightiest elephant-hunter of these or any other times, who has devoted the greater part of the last seventeen years to the fierce toil and countless dangers of ivory-hunting, and the numerous Dutch and other hunters who have ravaged the country, have much to answer for in this matter. But, after all, they have been but the instruments of supply to a pitiless and never-ceasing demand.

Turning westward to Bechuanaland, the same story has to be told. In Gordon Cumming's time— 1846 to 1850—Bamangwato, now ruled by Khama, was a veldt virgin to the hunter. Gordon Cumming himself was one of the first to exploit it; and by him, Oswell, and Vardon, and afterwards by Baldwin and others, great execution was wrought amongst the elephants, even as far to the west as Lake N'Gami. Professional hunters have, since

Gordon Cumming's day, completed, absolutely, the work of extirpation in these lands. In one year alone, after the discovery of Lake N'Gami (1849), Livingstone tells us that 900 elephants were slain in the region of the Great Lake. What wealth of animal life, although backed by countless years of undisturbed freedom and repose, could withstand the barbarous ravages of so short-sighted a policy? North Bechuanaland, including its Zambesi borders, and the Mababé veldt—once a great hunting ground —and the Lake N'Gami regions, are now all but completely denuded of ivory. Passing yet further to the westward, across the Kalahari, which now shares the dearth of Bechuanaland, we reach the countries of the Namaquas and Damaras. Here the same miserable history has to be recorded. Since the time of the explorer, Charles Andersson, between 1850 and 1860, when elephants were found in abundance in South-West Africa, succeeding years have seen the professional hunters pursuing without mercy or cessation their work of slaughter, and elephants may now be found no further south than Ovampoland, where the native hatred of the white man has alone protected them. The ivory trade of South Africa has, with the decline of elephant life, decayed in like ratio. In 1875, the value of ivory exported through Cape Colony was £60,402; in 1886, it was £2,150! In 1873, £17,199 worth of ivory was exported from Natal; in 1885, but £4,100 worth! Beyond these Austral-African regions, where ivory is now as scarce as it was but a few years since superabundant, the progress of extirpation goes on apace. Portuguese and Arab hunters have for years been hard at work, and each season sees the precious supply wax smaller. In the Kilimanjaro regions in East Africa, and about Bengweolo and the other great central lakes, elephants are, no doubt, still very abundant; but each month rifles and hunters penetrate more deeply into the Dark Continent. Probably the Niger territories, and the little-known interior of Western Africa, will hold out longest of all. But when it is remembered that 1,200,000 pounds weight of ivory are yearly imported into England, for which supply it has been computed that 50,000 elephants are annually slain; when it is remembered that for the annual supply of the whole world 100,000 elephants must yearly suffer death; no sane person can resist the conclusion, that the days of the African elephant are now " but as grass," and that his place will soon " know him no more."

Chapter XIII.
SPRINGBOK SHOOTING.

ONE fine morning, we left Naroekas Poort to visit a friend of our host, who farmed on the karroo, some twenty-five miles away, and who had invited us for a few days springbok shooting on the flats around his house. This we gladly accepted, for at that time we had only had an occasional shot at long ranges at these antelopes, as we drove across the Camdeboo Plains from Graaff Reinet to Naroekas. There were four of us—our host, H., who rode, and his wife, Bob (one of my travelling companions), and myself, who bestowed ourselves in a Cape cart drawn by a pair of horses. We trekked six or seven miles up some of the most villainous mountain roads to be found in the Colony, which is saying not a little. The day was glorious, and the scenery grand. Here and there a troop of baboons shambled alongside, at a respectful distance, barking angrily at our intrusion. Aloft, in the clear sky, some vultures circled, watching intently some object beneath them. Ringdoves, kingfishers, and honeybirds imparted life to the scene. After many shakings, bangs, and bruises, in which, I fear, Mrs. H. came off worst, we emerged from the mountains, through the wildly picturesque pass of Swanepoels Poort, into the plains. Here, after crossing the very nasty drift in the Plessis, a tributary of the Gamtoos or

Groote River, where the banks are deep and almost perpendicular, and where, on our way to Naroekas, we had nearly come to grief in an exceedingly dark night, we at length entered upon the eastern edge of the Great Karroo, upon the flat surface and improved roads of which we bowled merrily along. We had but one short outspan on the veldt for lunch, and then proceeding, arrived at Riet Fontein, our destination, in the afternoon. Here we were heartily welcomed, and after coffee we were shown round the very extensive kraals and farm buildings. Riet Fontein, like many another Karroo farm, lies shadeless far out on the heath-like plains, where are daily pastured many thousands of sheep and goats. The name of our entertainer—Mr. J. B. Evans,* " of Riet Fontein "—is so widely known throughout the Colony, as that of a successful and enterprising pastoral farmer, and one of the earliest introducers of Angora goats, that I may be excused for mentioning it. Mr. Evans, as is well known, has done, probably, more towards the successful introduction of Angoras than any other man, and he has even visited Asia Minor for the purpose of obtaining the purest strains of this breed. Having inspected the horses and ostriches, of which latter great numbers stalked solemnly about in their enclosed "camps," we strolled down to the dam to see the vast herds of sheep and goats come, in charge of Kaffir herd boys, for their evening drink, before being kraaled for the night. A beautiful and truly Eastern sight it was, to see the different flocks coming in in thousands from their long day on the

---

* Recently deceased, to the great regret of all who knew him.

thirsty veldt. After this, we adjourned to supper, and thence to Mr. Evans's sanctum, to indulge in pipes and a glass of "Cape smoke." During the evening, we arranged our plan of campaign against the springboks on the morrow, and it was finally settled that Bob and I, with Mr. Evans's son, Edgar, should rise before dawn, and get out on the veldt, among the bok, under cover of darkness. By so doing we should probably get a good shot or two before H., later on in the morning, came out to us with horses, when we were to try a "springbok drive." I may mention here that a herd of some 4,000 of these beautiful antelopes grazed upon Mr. Evans's veldt, some 256 square miles in extent, and numbers of them could be always seen grazing quietly within a few hundred yards of the house door. .The springbok, the ostrich, and the steinbok, are now the only representatives on the Great Karroo of the teeming herds of game that formerly abounded there. It is not so very many years since the gemsbok, wildebeest, hartebeest, ostrich, and quagga imparted life and beauty to these parched plains. In 1837, when Captain Cornwallis Harris (one of the most accomplished of South African hunters and naturalists) passed through the Colony, on his way to the country now called Transvaal, then a *terra incognita*, he met with the black wildebeest, quagga, and ostrich, in large numbers on the karroos not far from here.

In a curious old map, illustrating Barrow's travels, published in 1804, the karroo region is marked as abounding with the eland, gemsbok, hartebeest, quacha (quagga), wildebeest or gnu, ostrich, springbok, lion, and other animals. Most of these

HEAD OF A SPRINGBOK EWE (Profile)

animals have, however, been driven from their ancient plains; and the springbok, steinbok, jackal, and a few wild ostriches, alone remain. There is, however, an abundance of game birds of various kinds. Just as we shooters were about to retire, three uncouth and very dirty Boers came in to see our host on business. After their invariable custom, they walked solemnly round, and placed their flabby paws in everyone's hand, but without looking at one or speaking; they then sat down to a long and desultory conversation, and as at that time our knowledge of Boer Dutch was very limited, and the visitors appeared far from interesting, Bob, I, and Edgar shortly retired to rest, after looking to our rifles and other impedimenta. At about three o'clock we rose, dressed quickly, and stole quietly out. It was the spring of the year, the air was keen and frosty, for on the elevated karroo plateau frosts are severe, and a dense mist hung over the veldt.

Not a sound could be heard, not even the distant bleat of a springbok, or cry of a jackal. Edgar, who is accustomed to this sort of work, leads the way, and we walk across the veldt for half-an-hour, as briskly as the darkness, the unevenness of the ground, and the rough heathy scrub would permit, stumbling now and again as we go. Then our guide stops to listen for a few moments. We resume our silent march for another half-hour, when, having progressed some four miles, we again halt. This time Edgar whispers that he can hear the springbok, and again we proceed with renewed caution. In another ten minutes we stop again, and this time we can all distinguish sounds to our front, and an occasional bleat.

We hold a whispered council, and after waiting for a while steal forward very cautiously, keeping what little breeze there is in our faces. In ten minutes a subdued "hush!" brings us to a halt, and we sit down waiting for light. There is now at hand the faint approach of dawn, and shortly, as it lightens perceptibly, we rise, shove cartridges into our rifles, and peer anxiously through the mist. Then Edgar pushes Bob forward a few paces, pointing to an almost invisible object to his front; at the same time I can just distinguish something moving to the left, fifty yards away. In a few seconds, as the light comes, we can just make out the forms of three or four springbok grazing, and at a glance from Bob to indicate that he is ready, we simultaneously fire, Bob to his front, Edgar and I at an antelope on our left. Bob having a longer shot misses clean in the uncertain light and fog; but a loud thud follows the reports of our other shots, and the antelope we aimed at plunges forward and falls. Rushing up, we find a springbok struggling on the veldt with a bullet (Edgar's, as it turns out) through its shoulder. Its life is quickly ended, and then marking the spot with a handkerchief stuck on a stick brought for the purpose, we run briskly forward for a mile. Again we have to proceed slowly, as the fog shifts back again; but in a few minutes it clears, and Bob and I get another shot at a bok which trots slowly past us within thirty paces.

The beauty has not time to escape before we have each planted a bullet, one through the brain —a lucky shot—the other in the centre of the body, too far back to have been immediately fatal of itself. This bok falls dead to the first-named shot, and is at once secured and marked down.

The mist now began to lift rapidly, and, as the game around us was too disturbed to enable us to do much more good at present, we ate some biltong (dried meat) and bread, and took a soupje of Boer brandy; both welcome after a cold morning on the veldt. We had now nearly an hour's leisure before H. and the horses were to join us, and had time to look around as we strolled quietly in the direction of the farm. As the sun began to make his presence felt the mist rolled away, and shortly Aurora appeared in her most resplendent toilet.

That sunrise on the karroo I shall not readily forget. As the sun ascended in a perfect blaze of colour—pale blues and greens, and flaming reds and yellows—the vast expanse of open plains stretched in limitless extent before us, till broken by the dazzling horizon of the east. In every direction around us we could distinguish the herds of springbok grazing within a mile or two. The mountains we had quitted yesterday rose solidly in a semicircle behind us, springing from the plains in stern magnificence, and tinted lovingly by the rising orb. Presently, as the heat increased, the mirage danced before us, presenting to our enchanted vision the most curious effects. As we looked at Riet Fontein, some three miles distant, we saw, standing in air immediately above the farmhouse and buildings, an exact counterpart of them, but inverted, with the roof downwards. Turning in another direction, groups of trees and lakes of water appeared, startling in their semblance to reality, even to us, who knew that the brown karroo was destitute of any moisture except in the artificial dam at Riet Fontein. In another half-hour the mirage disappeared, and the most marvellous clearness of atmosphere succeeded. The tawny mountains, in no place nearer than twelve or fifteen miles, appeared almost within rifle-shot, and we could plainly distinguish every little kloof, crevice, and inequality sharply defined upon their rugged surface, until they melted, many a league away, into deepest blue. As we gazed again upon the open karroo, in the other direction, but one object interposed between us and the far horizon—a conical mountain (the Tiger Berg), rearing itself from the sea of plain, and looking down in solitary grandeur on the world beneath. The air, upon these elevated plateaux, possesses an extraordinarily exhilarating influence, and I think one never feels "fitter" than in these regions.

At length we perceived H. and a Kaffir riding towards us, the former leading one, and the latter two horses. As soon as they had reached us, we buckled on our spurs which they had brought, and despatching the Kaffir with the fifth horse to pick up the game we had shot, listened to H.'s instructions as to the best method of driving a herd of springbok in the distance. In the result, we mounted, spread out upon the plain at distances of some three hundred or four hundred yards apart, and walked our horses quietly in the direction of this herd, which numbered some three or four hundred antelopes. When we had got within three-quarters of a mile, the antelopes became disturbed, and began those extraordinary saltatory accomplishments ("pronken," the Boers term them), from which they take their name. One of the herd, followed by several others, would spring sheer and straight from its four feet, with arched back, ten or twelve feet into the air, as if made of india-rubber; this leap would be repeated half-a-dozen times or more, and then the animals would settle to a canter, and thence into a gallop. While these marvellous bounds are being executed, the springboks erect the curious mass of long snow-white hair, which extends from about the middle of the back as far as the tail, imparting a most singular effect. When the animal is not excited or alarmed, this hackle or ruff lies closely to the back, and is almost enveloped in the loose fawn-coloured skin which closes over it. Finally, the whole herd galloped away to a distant part of the veldt, long before we were near them. Disappointed in this direction, we turned our attention to a herd some distance to our right, and again attempted a drive, and this time with more success. This herd, some two hundred in number, allowed us to approach within seven hundred yards, and then H., who was on the extreme right, and had managed to creep outside and a little beyond them, suddenly began to gallop his hardest towards the game.

This was the signal for ourselves as well as the springbok. It is a well-ascertained fact, that the springbok, when pursued, will generally take up wind, and making for a certain point, they never swerve from their line. Immediately, therefore, we other three on the inside saw the antelopes crossing up-wind obliquely across our front, we set spurs to our nags, and raced our hardest to cut them off, or, at all events, a portion of them. Five minutes' desperate galloping across the rough veldt brought us a little before the bok, within forty or fifty yards of their line. No time in this scramble is there to look out for meercat or aard vark holes; we trust to our good horses and to Providence. Then we pull up suddenly, jump off our nags, throw the bridles over their heads, so as to hang in front of their forelegs (an invariable precaution against bolting, which every Cape hunting horse understands and obeys), and, cocking our rifles, are ready. In a few seconds, part of the herd come flying past at whirlwind speed within fifty yards. Crack go our three rifles, and one antelope drops, but regains its feet and limps after the rest. We hastily reload, and again the three barrels flash, but this time, in the excitement and at the speed the game is retreating, we all miss clean. The antelopes are soon far out of range, and scouring the veldt a mile or two away.

After some little trouble, and the expenditure of another cartridge, we bag the wounded bok, and then H. canters up. He is, and with reason, a little disappointed at our bad shooting, and tells us we ought to have bagged three or four head of game, but we remind him that two of us are tyros at the business.

We now rest a few minutes and give our horses a blow, and as the veldt is pretty well disturbed and the antelopes in the vicinity have nearly disappeared, we have time to admire the proportions of the dead game.

The springbok (*Gazella euchore*), tsepe of the Kaffirs, Matabeles, and Bechuanas, stands some thirty-two inches high at the shoulder, and two inches higher at the croup, and in length measures

HEAD OF A SPRINGBOK EWE

about fifty-eight inches. Its body colour is of a bright rich fawn, with a deep chestnut band running along either side from flank to shoulder, and slight bands of the same colour on either side of the white ruff or hackle I have mentioned as lying on its back. The head and face, throat, belly, tail, and inner parts of the limbs are white, rivalling in purity the snowy " blaze " formation on the croup. The horns are twelve or fourteen inches long, lyre-shaped, black, stout, and strongly annulated, and the tips bend inwards. In the female, the horns are slighter than in the male. The eyes are large, dark, and tender. The slender wiry limbs, and light well-proportioned frame, are fashioned exactly for speed and grace, and for those marvellous displays of strength and agility of which we had been witnesses. The springbok can exist for long intervals without water, and seems to have been designed by Nature to adorn the parched karroos over which it delights to scour. The flatness of its habitat enables it to scan attentively and suspiciously every approaching danger, and I suppose, for these reasons, the springbok has suffered (except in trek bokken or migrations) less than any other antelope from the attacks of the *Felidæ* and other wild animals. Altogether, the springbok, with its grace, its speed, and its agility, is one of the most beautiful and typical of the twenty-five varieties of South African antelopes, and standing as it does mid-way in size between the gigantic eland and the tiny blauwbok, is a worthy representative of the magnificent game that formerly crowded these vast plains.

We had noticed that the herds had now scattered far and wide, with the exception of a small group of half-a-dozen springbok that still grazed some half a mile distant, and as these fed quietly enough, I walked gently to within 500 yards, and then, lying on the veldt, took careful aim at the nearest. My aim was, rather to my surprise, true; the bullet clapped loudly, and the springboks, after a jump or two, flashed away at top speed. I noticed that the bok I had aimed at lagged in the rear, and determined to follow him; I ran back, therefore, for my horse, and found that my friends had had enough shooting for the present, and were off to breakfast. Mounting my nag, I told them I would not be long after them, and, amidst some chaff, rode away for a thickly-bushed kopje (a small hill or swelling in the veldt), towards which the wounded bok, which was now detached from its fellows, was retreating. I galloped hard, but the wounded antelope entered the covert before I could get within range. I soon reached the eminence and entered the bush; but quickly found it too thick and stony for riding, and, dismounting, left my horse. After wandering about some time, over rough ground and through thorny bush in search of the quarry, but without success, at length I gave up the search.

It was now ten o'clock—the sun was approaching hotly towards his zenith; I had been out since three o'clock a.m., and was parched and tired. The distant tinkle of goat bells, borne on the warm breeze across the sun-dried karroo, told me that the flocks had long since been unkraaled, and, altogether, I began to think my friends had not done unwisely in trekking for breakfast. Turning my footsteps, therefore, after some little trouble, I found my horse, who stood where I had left him, more patient by far than any human being. Good old Ombrāle! No better shooting horse was ever saddled on South African veldt.

Just as I was remounting, I noticed the old horse looking, with cocked ears, intently into the bush on his off-side. Looking that way also, I suddenly set eyes on a lovely little red antelope, not thirty yards away, staring fixedly at us with startled, yet inquisitive gaze. I lost not a moment, but taking rapid aim at the little beauty, planted a bullet just behind the shoulder. The steinbok, for such it was, fell dead to the shot, and with delight I rushed forward to secure it. After all, thought I, I shall have the laugh of my companions; so quickly disposing of the little bok behind my saddle, I descended the kopje, and cantered contentedly for home. The steinbok (*Tragulus rupestris*), eoolah of the Kaffirs, is one of the smallest and most beautiful of antelopes; its favourite habitat is just such bushy, stony hillocks, as that in which I had found it. It stands but twenty or twenty-two inches high, is about thirty-five inches in length, has sharp straight horns four inches long, beautiful eyes, and a lovely reddish-cinnamon coloured coat. Like the klipspringer, its pasterns are very rigid. The steinbok is found everywhere in the Cape Colony, and, like the springbok, affords good sport before the colonists' greyhounds. I was not much more than half-an-hour in reaching the farm, and found the others still busy at breakfast. I am bound to say all were pleased at my success. After a wash, I quickly joined them, and manfully performed my share in the demolition of springbok fry and chops, honey, coffee, and other good things.

I found that H. had shot a koorhaan (a fine game bird of the bustard species) with a single bullet as it stood on an ant-heap; and I was further informed, amid roars of laughter, how Bob, whose horse had put its foot in a meercat hole, had described an arc with unerring precision in his passage through the air to mother earth. While we were at breakfast, a pretty little fawn of the vaal or grey rhebok, which had been caught in the adjacent mountains, trotted about the room as tamely as a domestic cat, and evidently fully appreciated such luxuries as bread and sugar. The fawns of this antelope, and more commonly still of the springbok, are often taken when young, and make pretty gentle pets.

After a memorable breakfast, we adjourned to the stoep or verandah for a smoke, and a chat upon sport and antelopes. Mr. Evans told us of his experience of the last great trek bokken (migration) of the springbok in the Colony, which happened, I think, in 1858. In old days, these trek bokken were a source of the greatest alarm and danger to the colonist; quite as much, in fact, as the locust flights. Countless thousands of these antelopes, impelled by drought and the loss of their more secluded pastures, migrated from their true nursery and headquarters in the country formerly known as Great Bushmanland, now forming the districts of Namaqualand, Calvinia, Fraserburg, Clanwilliam, and Victoria West, and even from the far Kalahari Desert itself, where also they abound in vast numbers, into more fertile districts in the interior of the Colony. A trek bokken might be witnessed for a whole day, and the veldt would be left denuded of every scrap of pasturage. The immense numbers of the antelopes literally swept everything before them, and farmers frequently lost whole flocks in consequence. Our host described the approach of the trek bokken I speak of; enveloped in clouds of dust the herds came on. At one time the sight was positively alarming, for the springbok, on these occasions, from sheer press of numbers cannot retreat, and one has to be careful to keep out of their way. As the leading antelopes feed and become satiated, they fall back and allow those in the rear to come to the front; but for this provision of instinctive nature, the rear-guard would be starved to death, for those in front, of course, leave not a particle of nutriment as they pass. During these migrations the farmers shoot as much venison as they desire, and prepare immense quantities of biltong (salted and sun-dried flesh), of which the springbok furnishes the best quality. Mr. Evans informed us that during this last great trek bokken, he killed with buckshot no less than five bok at one shot. Of course, he was almost within touching distance, and the heads of the antelopes were close together; for on these occasions they are wedged so tightly that escape is impossible; and, indeed, it is actually on record, that lions have been carried along *nolens volens* in the midst of a trek bokken.

At the present time, the springbok is still fairly abundant on the Great Karroo and other plains in the North of Cape Colony; but in the remote colonial districts, formerly called Great Bushmanland, they are yet found in very large numbers, and there the trek bokken, on a smaller scale than of old, still continues. The Great Karroo is, however, so rapidly becoming more settled, and wells and water-mills, and the enclosure of pastoral farms are, under the improved conditions of farming, coming so much more into vogue, that I fear the springbok will, not many years hence, be driven to its strongholds and sanctuaries in the arid regions of the north-west of the Colony. There the country is so vast, so little settled, and so waterless, that many years must elapse before this beautiful and characteristic South African antelope is finally exterminated within the Colony. Even then there will remain beyond the Orange River the vast solitudes of the Kalahari Desert, where the springbok teems in countless numbers. Pringle, in one of his poems, says:

> O'er the brown karroo the bleating cry
> Of the springbok's fawn sounds plaintively.

And I am inclined to think and hope that the springbok's bleat will be heard on South African veldt long after most of the other antelopes are extinct. From its fecundity, its fleetness, its timidity, and extreme wariness, it seems probable that this antelope will decorate for another century the sun-dried pastures whereon it has disported itself for, probably, some thousands of years.

Chapter XIV.

A KARROO FARM.

THE morning star of Africa has some time since sprung like a rocket above the horizon; it is now approaching six o'clock, and after the usual early coffee, we are up and out of doors. Already the sun, although not long risen above the plains, is shining with a brilliancy known only in the marvellously translucent atmosphere of Southern Africa. The bold range of Witteberg, here forming the extreme south-eastern barrier of the Great Karroo, springs from its bed of mighty plain in everlasting, ever-imposing grandeur. The rosy flush, cast by the rising orb upon the brown and rugged bosom of the mountain, has not yet departed, and we stand for a few minutes watching the glorious tints gradually fade, leaving the mass of mountain, clad in its soberer, but little less beautiful garb of browns and purples. The place we are staying at is so good an example of a large pastoral farm in Cape Colony, that I may be pardoned for giving some particulars concerning it. Riet Fontein (reed fountain), then, lies to the extreme south of Camdeboo, an old Hottentot region, at the south-eastern angle of the Great Karroo, and to the south-west of Zwart Ruggens, another district, partly karroo, partly mountain. It is pitched upon the open plains at three or four miles distance from the Witteberg range, which, by the way, in this clear atmosphere, usually appears not much more than a few hundred yards away.

A ride of an hour or less up constantly rising ground, brings one to Schoorstenberg (chimney mountain), a bold and isolated pile of rock standing solitary upon the veldt, and rising upon its northern side (which presents a curious resemblance to a chimney) to a sheer elevation of more than 5,000 feet above sea level. From this peak, on a clear day, a magnificent panorama greets the eye. If you turn to the north-west, you may stretch your gaze across the broad expanse of plains until arrested by the Beaufort West mountains, which form the northern limit of the karroo, some eighty miles away. To the north-east, the bold range of Fore Sneeuwberg, beneath which shelters the town of Graaff Reinet, billows blue upon the skyline, 120 miles distant. South-eastward, the magnificent peaks of the Cockscomb — the highest point, nearly 7,000 feet in altitude, of the Great Winterhoek — rear aloft, plainly apparent, though ninety miles away, from amidst a tossing sea of mountain; while, to the south-west, the Zwartberg range may be readily picked out in the neighbourhood of Prince Albert, although not less than 180 miles distant.

Yet one other natural landmark strikes the eye. This is the Tiger Berg (leopard mountain), a conical peak, thrusting itself 3,000 feet sheer from the karroo, some twenty-four miles distant from where we stand. But although Schoorstenberg is, from its isolated position, a sufficiently striking object, it is not equal in height to portions of the Witteberg range lying immediately behind it. The highest point of Witteberg nearest Riet Fontein, is at De Beer Vlei, where an altitude of 6,000 feet above sea level is reached. Thus, at Riet Fontein, the dull monotony of plain is relieved by the noble mountains rising here and there around.

I believe most people, who have not been there, picture to themselves South Africa as a series of flat and sandy plains. Nothing can be more erroneous. The surface of the whole country is, on the contrary, ribbed and broken with mountains, between which here and there lie plains, sometimes, it is true, of vast extent, as in the case of the Great Karroo. The karroo is, in fact, the largest of the South African plateaux, having a length from east to west of 350 miles, with a breadth of from seventy to eighty miles. Nothing can be more magnificent, nothing more inspiring, than a journey through these mountain interiors, especially when the traveller emerges suddenly through some frowning poort upon the broad and limitless expanse of the Great Karroo.

In the year 1860, our host, Mr. J. B. Evans, first came here, and purchased the farm of Riet Fontein from an old-fashioned Boer, one Gert Hendrik Stols. At that time the place was little better than open desert. There was a house, certainly, the nucleus of the present extensive buildings; there was one very small dam, and there were a few insignificant fountains rising in the neighbouring heights of Witteberg and Schoorstenberg, which sufficed for the watering of Stols's flocks—that is, if the seasons were not too dry, or the drought too prolonged. But what changes have taken place since 1860 under the management of an energetic Welshman. In place of the ramshackle homestead of the old Boer, Riet Fontein* is now almost a village, and boasts a population of some sixty souls —whites and blacks—while upon the whole run there are nearly 100 people more. Farm has been added to farm, until the total area of the estate contains not less than 164,000 acres of mountain and plain, of which 132,000 acres are karroo, and 32,000 acres mountain pasture. Two hundred and fifty-six square miles of land seems an unconscionable deal of space; but large flocks demand far-spreading pasturage, at all events under present conditions of the Cape.

The Cape Dutch distinguish pastoral farmers under two heads, viz.: Groote-Vee Boers, *i.e.*, great-stock farmers; and Klein-Vee Boers, little-stock farmers. A Groote-Vee Boer runs horned cattle, whilst a Klein-Vee Boer depastures sheep and goats. Our host ranges himself under the second denomination, and is a Klein-Vee Boer. Upon his 164,000 acres are depastured 15,000 sheep, 20,000 goats, and 500 ostriches, whilst, in addition, 200 head of cattle and 100 horses are run upon the slopes of Witteberg and Schoorstenberg.

These flocks are, of course, not all quartered at Riet Fontein itself; the farm is divided into twenty-five out-stations or vee-kraals, scattered about in different parts. Each of these vee-kraals is in charge of a Dutch or native overseer, under whom are numerous herds—Kaffirs, Bushmen, Hottentots, Bechuanas, Mantatees, and others— who daily tend their flocks at pasture, and bring

---

* NOTE.—I am describing the farm as I saw it a few years back. It was sold by Mr Evans, subsequently, and is now the property of the Cape Stock Farming Company.

them home at night to kraal. Many of these out-stations lie still farther out into the plains, and have not all dams or fountains upon them. The flocks, however, get water once in three days in the cooler season, in this way:—It is arranged that they shall depasture on one side of the station for one day without water; the next day they are fed upon the other side of the station where water lies. Here they drink, return to kraal as usual, and next day are depastured in another direction without water. Thus one dam suffices for three, or even more out-stations; that is in fairly cool weather. When the season gets hot, they require water every other day, and at the homestead, where water is abundant, they drink night and morning. From April till September—the Cape winter—sheep and goats alike, are often entirely without water, unless, of course, near an abundant supply; in this cool season they scarcely need it, and obtain a sufficiency of moisture from such succulent plants as the spekboom, braak veije, a small kind of stunted aloe called alveig, and others. But, occasionally, the karroo is visited by prolonged and appalling droughts, when but little rain or moisture falls for two years at a stretch; then every dam dries up, every atom of moisture is scorched from the herbage, the plains gasp under clear and brazen skies, and the flocks perish in thousands. During one of these droughts, our host himself lost 20,000 head of sheep and goats, while farmers, not so well provided with water, lost even more heavily in proportion.

Where the veldt is good, sheep and goats can, if required, during the winter and spring months—

from April till October—subsist entirely without water. The karroo vegetation, despite its parched appearance, is so succulent, so full of moisture, that unless near a homestead dam or fountain where water can almost always be spared, the flocks can without the least inconvenience do entirely without water. It may be astounding, indeed, to the unitiated to learn that in South Africa, of all places in the world, animals can thus live and thrive for long months upon pastures absolutely waterless. Yet such is the case. Some years ago, when the veldt was virgin and untrampled, our host kept his flocks upon the open karroo, on the Aberdeen Road, for the greater part of two whole seasons, without giving them any water whatever; nor did he find that they suffered in consequence. There was one small fountain on the run, just sufficing for himself and his horses, but no more. During the extreme heat of summer, the flocks were removed to other pastures possessing water. But at the present day, it is impossible to find on the Karroo such fresh and virgin feeding grounds as then existed. There has been too much overstocking; and far too little care has been taken to preserve and give rest to portions of the runs. The system of wire fencing, however, now becoming common, will soon work wonders in this respect.

Riet Fontein itself stands close to a dry karroo river-bed, which holds water only in time of heavy rain, and then but for a short period. As with most South African rivers, there is a heavy flood which abates as rapidly as it rises, and the river-bed then resumes its normal appearance. Here and there, indeed, perhaps within three miles or so, there is a

pool where the water escapes from underground from the dykes higher up. These pools are often brackish, and become undrinkable as drought increases. The river-bed is handsomely margined with a fringe of mimosa trees (*Acacia horrida*), which mark out plainly for miles along the flat expanse its devious and meandering course. The old Boer house has been enlarged and added to, until there is now an irregular pile of buildings of red Kaffir bricks, concreted with lime and sand—rough casting, as it is called—covering a considerable space of ground. Here dwell our host and hostess and their olive branches, and their tutelary deities. The nearest good school is, perhaps, 150 miles away; and instruction, thorough and systematic, is carried on upon the premises under a careful and accomplished tutor.

At the front of the dwelling house runs the invariable Cape stoep or raised verandah, which offers a cool and refreshing shade to weary souls. This stoep is partially covered with a handsome vine. Fifty yards in front of the homestead lies a good-sized dam, and in the rear a much larger one, constructed at great cost and with infinite pains. In addition to this main dam, there are, as I have said, others at something like half the out-stations. But besides dam-making on a very large scale, our host, after experiencing disastrous droughts, when even the dams became exhausted by evaporation and by ever-thirsty flocks, turned his ideas to other methods of procuring water. He was the first farmer, I believe, to bore for water beneath the karroo, and now, at a depth of 212 feet, he procures through a two-inch pipe a sufficient supply of the precious element to water, if need be, from 8,000 to 10,000 sheep daily. Each sheep drinks in summer about one gallon at a time, so that the supply from this source may be taken as equalling nearly 10,000 gallons daily. Round the margin of the dam are planted graceful willows, fig trees — which offer delicious fruit in season—and blue gums; and upon a well-watered patch of the rich red karroo soil hard by, flourish in perfection tomatoes, potatoes, cabbages, turnips, lettuces, and radishes. To the left of the great dam, and at a good distance from the house, stand two sets of the kraals, wherein are gathered nightly the flocks of sheep and goats on their return from the home veldt. These are the oldest of the kraals, and are built of the old-fashioned Boer bricks, made of dung cut in blocks from the hard-trampled enclosures, and dried hard and clean in the sun. This substance, locally called "mist," serves also almost entirely for fuel on karroo farms, where wood is scarce and expensive; it makes excellent firing burning slowly much as peat does. The kraals are wisely placed at a respectable distance from the house. The Boers usually build theirs close to their dwellings, and are in consequence in summer plagued to death with flies. These enclosures are partitioned off into sub-divisions.

Crossing the dry river-bed from the open karroo side, whereon our host's house stands, we come first to the comfortable roomy dwelling and stables of his partner, Mr. John Rex. Some distance to the rear of this house, partly upon the rising ground of a kopje covered with Kaffir plum, spekboom and other bushes, partly upon open flats, lie the ostrich camps—immense enclosures, girt by wire-fencing, wherein 500 birds can wander freely. On the left hand, as we pass Mr. Rex's, lie the houses of the blacksmith and handy man—a very important personage here—and other workpeople, and of the herdsmen; beyond, there are more stone kraals, wherein the remainder of the flocks are shut up for the night. Still further behind all these, forming a glorious and ever-welcome background, stand the mountains, stern and beautiful. But the flocks, to the number of 9,000 or 10,000, are about to be unkraaled, and we hasten to the oldest of the enclosures to watch the process. Standing by the gate is our host, who is desperately pre-occupied counting the Angora goats as they defile forth for their day's pasture on the veldt. This fleecy census is a business never omitted by the Cape farmer, and it is astonishing what marvellous rapidity and exactitude are attained. Mr. Evans is famous for his Angoras, and the beautiful creatures, with their long snow-white silky hair, are well worthy of the care and trouble bestowed upon their rearing. First brought into prominence at the Cape in 1856, Angoras have proved wonderfully successful, and the mohair clipped from Cape stock now yields little, if at all to the purest breeds of Asia Minor.

Our host, who has perhaps done more for Angora goat farming than any other man in the Colony, was one of the earliest to recognise their value. He has even undertaken a journey to Asia Minor in search of the choicest strains of his favourites; and from the districts of Tcherkess and Geredeh, by permission of the Sultan, he obtained some most valuable stock. In this expedition he undertook a journey of 1,200 miles inland, and brought out from an almost unknown district thirty or forty Angoras, which were carried to the coast packed on the backs of mules. Some of the rams thus imported realised from £100 to £400 a-piece amongst the farmers of Graaff Reinet and other parts of the Eastern province. But the Angoras have at length all poured forth from their kraals, and their silken coats glinting beneath the sunshine, are off to the broad plains for the long hot day in charge of Kaffir herd boys. Well may our host look lovingly after them. His clip of Angora hair is the largest in the world, and he possesses the greatest number of goats—20,000, all Angoras, or half-bred Angoras—to be found on one farm in South Africa. The annual clip of mohair from these flocks amounts to 50,000 lbs. Some farmers clip their goats twice a year; but this is a greedy and short-sighted policy, which in the long run brings weakness and deterioration alike to goats and to mohair. Upon the estate of Riet Fontein each year 4,000 kids are reared; occasionally seven hundred are dropped in one day. This may continue for a day or two, and then the number diminishes, perhaps to increase later, for, unlike sheep, the kidding appears to progress by fits and starts. During this season, the goats are kept in large camps enclosed by wire-fencing, and the kids are immediately after birth fastened to the fence by the hind leg. Goats resemble the springbok and other game in their propensity for leaving their young in one place and wandering away for a time. At one time the kids were fastened to the karroo bushes, but it was found that in looking them up and putting them into kraal, after fourteen days or so, many were hidden in the thick herbage and left behind; and

FLOCK OF 1,800 ANGORA KIDS.

the ewes, after a couple of day's absence, failed to recognise them, thus causing trouble and inconvenience. Now, when fastened to the wire-fencing, the herdsman simply walks up and down, and finds his charges at once.

The time for clipping the Angoras comes in July, when from fifteen to twenty hands are hard at work for three weeks or a month. The goats are clipped in forks, into which their heads are inserted, and in which they stand firm and rigid as rocks until the operation is completed. The silly sheep, on the contrary, cannot be induced to stand in this manner, nor perhaps, if they were willing, could the old-fashioned method of shearing be improved upon in their case. The yoke, in which the goats stand, is an exceedingly simple yet ingenious contrivance of our hosts, and deserves description. Two iron forks are fixed into a wooden standard, sufficiently high to reach to the goat's neck, the standard itself being firmly planted in the ground. One of these forks is movable, and is so contrived that the goat's head being inserted, the fork can be pushed back into its natural position, and a wire being then passed over the neck and fastened to both forks, the goat is firmly secured and cannot withdraw its head.

The 15,000 sheep depastured on the estate are almost entirely of the merino breed. The clip of wool from the flocks in one year amounts to 150 bales; that is, reckoning the bale at 450 lbs., 27,000 lbs. weight. Wool shearing begins in November, and some twenty hands, including sorters and packers, are kept busily employed for three weeks or a little longer. At the time I speak of, the wool or mohair when sheared was put on ox-waggons and sent to the Port Elizabeth market; now, there is a railway station at Klipplaat, within thirty miles of the farm, and the process is thereby expedited very materially.

The system of nightly kraaling the flocks, which even to this hour obtains to a very large extent over the whole of Cape Colony, is a survival of those wild and primitive days when the early Boers had to defend their fleecy charges from the lion and leopard, the various hyenas, the Cape hunting dog (*Lycaon pictus*) and jackals, and the smaller of the Felidæ and Viverræ innumerable. But of all these destroyers, the jackals, the African lynx, a hyena or two, and certain of the wild cats, alone remain to the karroo plains; the rest have been shot or poisoned off, or driven northwards to the interior. The leopard, it is true, still haunts the mountain ranges, but even he is being surely, if slowly, exterminated, just as is the jackal from the karroos.

With the disappearance of these carnivora, the advent of wire-fencing on an enormous scale has come. Huge runs are being completely girdled by wire-fencing, which is rendered proof against jackals and other vermin. Our host, whose complete and perfect system of wire-fencing now patented and shown in England, after years of patient and practical working out in the Colony, was one of the first to recognise its utility. Since the time I write of, Mr. Evans has fenced his entire estate with some fifty miles of wire-fencing, and the Cape Stock Farming Company, to whom he has sold the run, have now a fine property, wherein their flocks can pasture by day and night without being brought into kraal. The advantages of this system are manifest and manifold. Under the old plan, obtaining at the time of my visit, the flocks spent a good part of the day in journeying backwards and forwards to kraal, in the hot summer weather a matter of much fatigue and distress; thus from the perpetual trampling to and fro of mighty flocks, vast portions of the veldt were completely destroyed. Much of the manure which might have added to the fertility of the soil was wasted in the kraals, or only partially utilized in its sun-dried form as fuel. By these weary perambulations to and fro, paths were hollowed out, which increased in process of years into ravines, down which, in time of rainfall, the water poured away and was wasted, instead of being distributed equally over the surface of the soil. It is not too much to say that wire-fencing, now being adopted by all progressive farmers at the Cape, will effect immense saving, not only by the rest afforded to the pastures, but by the increased feeding and repose, and, consequently, much-improved condition thus afforded to the sheep and goats.

But we have at length seen every kraal emptied of its hungry contents, and have watched the stream of sheep and goat-life pouring out far upon the plains.

Breakfast is now announced, and with healthy appetites, sharpened by the clear bracing air of this elevated plateau—here some 3,000 feet above sea level—we follow our host to his comfortable dining room. Before us is spread a typical Cape breakfast; here is springbok fry (karroo grown, though it is more delicate and more savoury than that of the tenderest lamb ever nurtured in verdant

England), cold springbok haunch, cold koorhaan—one of the bustards—steaming coffee, goat's milk, delicious hot "cookies" of Boer meal, springbok biltong, sliced into the most delicately thin shavings by the carpenter's plane. Biltong, especially of the springbok (*i.e.*, flesh slightly salted and dried to the consistency almost of horn), when sliced in this way forms a most appetising relish, and is a most nourishing commissariat article keeping, although uncooked, always fresh. The Boers, in their hunting expeditions and their war commandos and forays, invariably use it, for it occupies little space, does not require cooking, and is always ready for eating. Then there is store of fragrant wild honey, taken from the neighbouring rocks of Witteberg, and to crown the repast luscious stewed peaches, with plenty of sweet goat's milk to wash them down; tomatoes and lettuces impart an air of Arcadian grace to the board. Talk to me of the discomforts and hardships of the Cape! With a little trouble, a little forethought, many a delicacy now unknown to the farmer's table might give zest to many a jaded palate. What a difference, this snowy cloth and bountiful table, spread though it is in the wilderness, to the filthy board, the everlasting greasy mutton swimming in sheep's tail fat, of the average up-country Boer. How we enjoyed that meal! Memories of it, tender and fragrant as aromatic karroo shrubs after the rain showers, still linger with me.

Breakfast over, and the little tame vaal rhebok, that trots so gracefully round the table pushing its soft moist muzzle into one hand and another, propitiated with offerings of bread and sugar, we light our pipes, and lounge for just ten minutes on the stoep (verandah).

Away out upon the plains yonder, the nearest of them not 700 yards distant, graze the dainty springboks, which yesterday we hunted. If we lay down and took a long steady aim, with carefully-adjusted sights, we might even knock one over from the door here. But it is neither fair nor discreet to disturb the beauties too often, and we leave them to-day to wander undisturbed.

Our busy host is soon ready for us, and at once carries us off to the ostrich camp to see some of his birds clipped. Passing a young half-tame baboon—known as Adonis—tied up to an outhouse, and just stopping to exchange a morning greeting with the cunning and amusing rascal, we cross the deep spruit of the dry river, and shortly arrive at the ostrich camp. This camp is a vast enclosure of between 3,000 and 4,000 acres, surrounded by wire-fencing about four feet ten inches high. Great care has been taken with this fencing, and it is now, by an ingenious arrangement of wires laid flat upon the ground, rendered jackal proof. These wires, twelve in number, are towards the bottom set three inches apart, the lowest one resting on the ground. The jackal begins to burrow, and expecting to get quickly beneath the upright fence, comes in contact with the bottom ground-laid wire. Instead of retreating a little and then beginning to burrow, the foolish animal (in other matters terrestrial crafty enough) runs farther down the fence, and tries again with same result; and, finally, after repeated attempts, throws up the sponge, and admits the fence to be the conqueror. This is a well-ascertained fact, and a very important one it is likely to be to the Cape farmer; although, so far I believe Mr. Evans is alone in the use of "jackal wire." The camp is partly flat karroo, partly rising and broken ground (kopje), whereon wilde pruim, spekboom, and other succulent bushes and shrubs flourish in a scant luxuriance. Within this huge enclosure or enclosures—for there are smaller ones carved out of the main camp—some 500 ostriches are gathered.

Here the great birds wander practically free and unfettered—some full grown, some smaller, others quite young—in solemn stateliness; ludicrous creatures they are with their quaint bald bristly heads. We enter the largest enclosure to assist in driving some of the birds into a small pen or fold between the great and little camps. A ferocious old cock comes rushing up with direful eye and malevolent intent, but is easily kept at bay by thrusting the long-forked sticks with which we are armed against his neck; this is a simple yet infallible guard, without which no ostrich camp should be entered if it contains dangerous birds; a thorn bush held in front of the bird's neck will answer the purpose equally well. A fierce old cock bird is indeed no joke, and can snap the limbs of the strongest man with a kick impelled by the enormous muscular power of his thigh. The old cock presently turns away in a huff, and stalks off to another part of the camp. Rampageous and unreasonable creatures as ostriches may appear to the uninitiated, they are in reality very easily managed. When they have to be driven considerable distances from one place to another, they are sometimes guided by a kind of harness. A band of leather is passed round the chest and under the wings; to this two long "reims" (skin halters) are attached, and the bird is then easily driven by two men—each holding a "reim." Another plan often adopted, is to fasten a "reim" round the bird's neck; this is held by a man on horseback, who canters alongside of the ostrich, and gets over the ground with his willing captive in very easy fashion. The ostrich is wonderfully amenable to pressure upon the neck, and makes the journey under such circumstances with far more than the docility even of the tractable sheep. After some trouble, a certain number of birds, which have been brought up early in the morning from distant parts of the camp by the Kaffir herds, are driven one by one into the small fold. Here they are guided into a corner, and then a hurdle is set up against them; then a bag or small sack is placed over the head; the bird stands perfectly still, and is speedily denuded of its best feathers by the aid of a pair of garden pruning scissors, which do their work neatly and expeditiously, leaving about half-an-inch of quill stump remaining; this soon afterwards becomes loose, and is easily and painlessly removed, giving place to a fresh growth.

There is no pain whatever in this clipping process, but in the earlier days of ostrich farming, when the feathers were pulled out bodily, I fancy there must have been some suffering. In many cases, instead of using bags for the head, boxes are employed during the process of denudation. One after another the birds are rapidly driven in, clipped, and turned out again, until there is a great pile of feathers — prime whites, feminas, drabs, blacks, spadonas, and the rest. At the time I write of, feathers had not yet fallen to the wretchedly low prices they now command, nor had the birds themselves lost much of the fictitious value they assumed during the extremest rage of the ostrich mania. About that time, £1,000 had actually been paid for a pair of breeding birds with a nest of eighteen eggs! £500 was not by any means an uncommon price for a pair of really good breeders; while from £200 to £300 were common enough prices. The sum of £82 has been known to be realised from the plucking of a few feathers only. But what a crash was there when the bubble burst, and the inflation was at an end. The Colony has not yet overgot its effects, which will be remembered with soreness for many a long day. Now (1889) birds are almost unsaleable, and good breeding birds may be bought for £5 a pair, and chicks for a few shillings each, instead of from £7 10s. to £10, as soon as they emerged from the eggs. Now, indeed, is the time to go into ostrich farming, for at the present low price of stock, it can assuredly be made to pay.

It has been commonly supposed that the domestication of ostriches began in 1864, from which time ostrich farming practically takes its rise; such is not the case. Kolben, that quaint and entertaining traveller, distinctly mentions having seen ostriches tame and in confinement at the Cape in or about 1705. It would seem, however, that the practice fell into obscurity, to be revived again with astounding and unexpected success 160 years later. It was a fortunate thing for these great birds that domestication set in at this later period, or a few years would probably have seen their complete extinction within the limits of the Cape Colony. The remnant of wild birds had by 1864 been greatly thinned by remorseless feather hunters, and ostriches were only to be found upon the Great Karroo and in Bushmanland and other remote parts of the Colony, where their young were often reared with extreme difficulty. Shooting parties were periodically got up by our host, and from eight to ten birds were usually accounted for during the hunt. Mr. Evans was one of the first three farmers in the Colony to inaugurate ostrich farming, the others being Mr. Kennear, resident magistrate, of Beaufort West, and Mr. Murray, of Colesberg.

The first chicks were obtained for him by a Boer, a beiwoner (a sort of sub-farmer on the estate of a richer farmer, who is expected to perform certain duties for the privilege of running his stock) on a neighbouring run. A nest was watched by this Dutchman until the chicks were hatched out; but when the time came for delivery, the Boer, who had had his suspicions, demanded much more than the agreed price—a trifling one—and obtained £4 10s. for half of the chicks, the rest going to the owner of the land. From these four or five chicks sprang, practically, the gigantic ostrich farming business that now obtains. At first all sorts of erroneous ideas prevailed concerning the new stock. As the birds did not breed for three or four seasons, it was imagined that the pristine wildness and timidity of their nature would prove a bar to their perpetuation while in captivity. Time proved this to be a mistake. Mr. Evans's first ostriches were not confined in camps, but ranged free upon the karroo, returning each night to kraal to be shut up.

As the cocks grew older they grew also bolder, and many ludicrous anecdotes of their escapades were told us. Two Dutchmen's horses, which had never seen an ostrich at close quarters, were nearly frightened out of their skins, while standing by the homestead, by the apparition of the cocks rushing upon them from round the corner of the house. Breaking their skin halters, they fled precipitately for miles across the plains, with the cocks in hot pursuit. Horses now view ostriches with as much unconcern as poultry. A Dutchman's horse was attacked from behind by the downward kick of an ostrich; the Boer was thrown and seriously hurt, and his gun broken, and an action for damages was nearly resulting. After these and other similar events, and as the cocks became daily more troublesome, they were hobbled for some considerable time, until the system of confining them in camps came into vogue. In fighting, ostriches very much resemble dunghill cocks. The two I have spoken of took up different ranging grounds, one selecting the higher, the other the lower part of the veldt. If he of the plains ventured into the upper ground to do battle, his defeat was certain; and if he of the upland sought his rival on his native heath, he was equally certain to be worsted; each bird invariably proving the *baas-raake* (conqueror) on his own territory.

One of the most singular habits of these singular birds is the challenge; at first this habit was quite incomprehensible to the early ostrich farmers. A cock would come rushing across the karroo in pursuit of a man on horseback, and when within a few yards, would suddenly bring itself to the ground, thrust up its head slightly backwards, set up its tail, and then, flapping its wings, utter a monotonous groaning chant; this challenge to mortal combat would be repeated over and over again. At first it was imagined that the ostriches were in convulsions or attacked by fits, but this idea soon faded as the birds persistently got to their feet again, assumed the defensive and displayed their fighting powers.

Another singularity about this bird is, that a cock, which has been dangerously pugnacious in its own camp, will, when removed elsewhere, invariably prove as quiet and submissive as the proverbial lamb. One other point. It is commonly thought that ostriches are monogamous; as a fact, they are only monogamous so far as the hatching-out is concerned. A number of hens lay their eggs in one nest; but the strongest hen then takes possession, and with the help of the cock, performs the hatching-out process. The younger and weaker hens often attempt to satisfy their "broody" instinct, by sitting religiously upon the bare sand as near the nest as possible.

Again, it has been thought that a certain number of the eggs are rolled out of the nest, and broken by the old birds to serve as food for the chicks as they hatch out. In reality, all those eggs that the sitting bird cannot properly cover and hatch are rolled out, and are then broken and devoured by jackals and other vermin. Nature's unerring instinct tells the birds that it is useless to retain in their nest eggs that cannot be hatched, and the encumbrances are at once got rid of.

At the present time, wild birds, from being strictly preserved, are much more plentiful; and, indeed, their numbers have been actually reinforced from the ranks of their domesticated brethren. A year or two back (1886-7) farmers became so utterly sick and disgusted at the fall of prices and the depreciation of their stock, that, rather than bear the expense and trouble of keeping useless and unremunerative encumbrances, they turned loose their birds upon their native karroos, where no doubt they will thrive and perpetuate themselves, and possibly impart to their wild congenitors new methods and increased store of artfulness—already possessed in plenty. It may be noted that "tame" feathers never equal the best of the produce of the wilderness; " Prime bloods," as the first quality of wild feathers are called, realising half as much again per pound as the best of domestic pluckings. The highest value in the export of feathers was reached in 1882, when 253,954 lbs. weight—valued at £1,093,989—were exported; in 1885, the figures had fallen to 251,084 lbs., valued at but £585,278—a terrible depreciation in value.

But now having done our duty by the ostriches, we are ready for the horses, which in the meantime have been saddled up and brought out to us by a couple of Kaffir boys. We are quickly mounted, and proceed to ride off to the slopes of Witteberg to have a look at some of the stock—horses and horned cattle—running there.

For part of the distance we ride across perfectly flat karroo, and our host tells us something about the nature of the herbage growing around. Just now we are nearing the end of a long droughty season; the early summer rains, so anxiously expected, have not arrived, and the far-stretching plains look too brown and barren to afford nourishment even to a rear-guard locust—an insect thankful for small mercies. Much of the red soil—rich enough for anything if water be brought to it—is scarcely covered; there are often great bare patches between the shrubs and plants, and the grasses have quite disappeared. But, droughty and parched though the plains appear to us after eight months without rain, what must they be like when, as is sometimes the case, no moisture falls for two years on end; two years of scorching suns, of brassy skies, of evaporating dams, and disappearing vegetation; towards the end, long months of weary struggling, to preserve the constantly decreasing flocks, which day by day die from thirst and starvation. At such times the lot of a Karroo farmer is not merely not an enviable one— it is an existence absolutely maddening. Yet even now, as we amble across the plains, there is in these apparently scrubby feeding grounds an ample pasturage, and the stock as we have seen are in good heart and fettle. These dried-up looking shrubs, and the low heathery brush that form the karroo vegetation, are even now full of feeding power; and when the rains come, as we afterwards witnessed, what a transformation! Brush and scrub, apparently devoid of life, shoot out a fresh and vernal verdure; starry flowers spring in profusion, even before the green leaves appear; fragrant grasses and herbs emerge as if by magic from the soil, and the whole surface of the karroo appears one immense ocean of dark green, spangled with flowers most brilliant and innumerable. A delicious aromatic odour fills the atmosphere. No one, indeed, who has not witnessed this glorious transformation, can possibly appreciate the vastness and completeness of the change.

Amongst the principal feeding plants upon these plains are the spekboom (fat-tree), guarray, karree-bosch, fingerpoll, vaal boschie, noorse-doorn (in certain localities, especially in Zwart Ruggens), and others too numerous to mention. The compositæ and the portulaciæ are very largely represented amongst the innumerable varieties of vegetable life. The hibis-cus, pelargonium, heliophila, oxalidæ (several), mesembryanthemum, crassula, and cotyledon, are very noticeable.

The fingerpoll is singular, even amongst many curious plants. This, the *Euphorbia caput-medusæ* of scientists, has a low growth, not distantly resembling a squabby footstool, with fingers standing forth from it. It is a tough, yet an extremely nourishing plant, and is so hard as to require to be cut up with an axe; the kernel or inner part of it is, when boiled, used as a vegetable occasionally. Sheep and goats feed greedily upon it when cut up and thrive amazingly. Oxen do not take to it quite so naturally, but when used to it they do even more credit to its recuperative powers; and it is a well understood fact, that spent and foundered oxen, left to die upon the road where they had fallen, have, when fed with fingerpoll, regained vitality, got up, and resumed their trek. Horned cattle, indeed, become extraordinarily fond of this plant. I have seen a span of oxen outspanned for evening, and wandering off into the bushes to graze. I have watched the driver go to the waggon, get out his axe, and with it smartly strike the wheels. The metallic clink was instantly noted and responded to by the cattle;

to them that welcome sound meant "fingerpoll," and briskly they all straightway returned and followed the driver a short distance, to where some of the much-loved plant grew. There they patiently waited while their provider cut up and prepared the luxury, and finally fell to with keen and hearty appetites, as their crisp crunchings plainly told.

We have a ride of several miles before us to the foot of yonder Witteberg; and accordingly, putting our nags to a canter, we rapidly get over the ground, rough and uneven, and studded here and there with meercat or ant-bear holes, though it is. The sun, now well overhead, shines warm and brilliantly; still it is not too hot, there is a pleasant breeze out, and the clear air is full of sparkle and exhilaration. Upon these high karroos, even in the fiery intensity of the hot season, life is not unbearable, and when

> "The sultry summer noon is past,
> And mellow evening comes at last,
> With a low and languid breeze,
> Fanning the mimosa trees."

the nights are cool, and sleep, welcome and refreshing to the sun-baked Vee-Boer, is obtained.

Far away to our right front there is a deep, dark indentation in the mountain range. That is Beer Vley, or De Beer's Vley, a pass from the karroo well-known to ancient travellers. Here Barrow rested and refreshed his cattle, and before him the lively Le Vaillant. Here amid the reeds and long grasses was slain, not so many years ago, the last lion known in this locality.

Presently, as we near the mountain and approached some thicker clumps of bush that begin to clothe the veldt, a sharp, fierce bark from the colley that accompanies us is heard in front. Then across a piece of open ground darts a reddish yellow form, followed by the dog. " Rooi cat! by jove!" exclaims our host; "shoot, or give me the gun, quick." But, alas! 'tis too late, the rooi cat has got to shelter, too dense to follow him into, and the colley is not equal, single-handed, to an encounter with an African lynx, and, indeed, is clearly not inclined for it either ; so after hunting and spooring about a little, we give it up. " We must have that fellow in the next few days," says our host ; "we'll bring the dogs and a Kaffir to-morrow, and hunt him up ; and if we can't get him that way, we must poison him somehow." The rooi kat (red cat) of the Boers, more commonly known to naturalists as the caracal or African lynx, is plentifully distributed over the Cape Colony, and as it is a sworn foe to the farmers' flocks is killed off as relentlessly as occasion admits.

A short way farther and the mountain is reached. We ride quietly about here and there, looking up and inspecting the horses and oxen—or most of them—which here run free and unconfined. When Mr. Evans, in 1860, bought Riet Fontein from Hendrik Stols, the old Boer had fifty mares, most of them with foals, which he grazed principally in the direction of these mountains. But so infested was the place with leopards at that time, that every one of these animals was knee-haltered, and they were all jealously tended throughout the day by sharp-eyed Bushmen or Hottentots. Every night they were driven into kraal, and every morning sent forth into the veldt again. But even with these precautions, the leopards secured many of the foals.

Our host has ended all this; strychnine placed in meat or dead carcasses, and set about on bushes or low trees, has done its deadly work; and on this side of Witteberg the "tigers," as they are invariably called at the Cape, now cease from troubling. Farther over on the other side, in Naroekas Poort and other parts of the mountain interior, however, they are still plentiful, and cause much damage. At Naroekas Poort, our mountain abiding-place, only two years before our arrival, the leopards had killed no less than fourteen foals in one season.

During the afternoon we inspect some eighty or ninety horses of all sorts, shapes, and colours, as well as a good number of oxen, and find them all pretty fit and flourishing. The Cape horse is, as a rule, a trifle undersized; is somewhat lacking in bone below the knee and hock, and too often has drooping quarters; yet, with these failings, I have found him (and most people who have tested the matter will agree with me) the hardiest and most willing servant in the world. No day is too long, no country too severe for him; and, in the matter of food, the Cape horse is as accommodating as such a-rough-and-ready Colony often requires him to be.

As hunting nags, when properly trained, they are the most sensible and sporting animals imaginable, and enter with extraordinary zest into the spirit of the chase; standing like rocks when dismounted, and galloping like the wind — carefully avoiding holes, with the cleverness of cats—when engaged in cutting off the flying herds of antelopes. Very many of the Cape horses have spotted or snowy quarters; I am not certain why this should be, but

I have heard that this characteristic is to be traced to their Barb or Persian ancestry.

The Cape horse owes its origin, some two hundred years back, to a Persian strain, and to ancestors imported from the Dutch settlement at Java; later on, to a touch of the Barb and a fair share of English blood, and the result, as I have pointed out, is an animal not by any means showy, but possessed of eminently useful qualities.

A pull at our cold tea bottles and a biscuit or two breaks our fast while we are shepherding the cattle. The slopes of these mountains are, in parts, very densely clothed with bush, the fleshy and succulent spekboom, and the strong wiry wilde pruimen, affording thick cover for game. Hither wander occasionally the magnificent koodoo (*Strepsiceros kudu*), most regal of South African antelopes, happily still to be found within Cape Colony, and, it is pleasant to say, even on the increase in some localities. Here is to be found, lurking in the darkest of the jungle, the bosch vark (bush hog), one of the two varieties of South African wild boar. Here the lovely steinbok, the fleet rheboks—red and grey—the crouching duyker, and the rock-loving klipspringers abound. But to-day we are not come prepared for shooting, although, as a matter of habit, guns are carried by two of our number; and as the sun now dismounts from his zenith, and the shadows lengthen, we turn our horses heads for Riet Fontein. A pleasant ride home, delayed a little by a fruitless attempt to stalk a steinbok, which is espied in a thinly-bushed kopje, brings us, towards the close of the afternoon, to the homestead, whose dam we can see shimmering beneath the slanting sunlight. We are not sorry to off-saddle, and have a cup of coffee, or something stronger, and a smoke during the hour before high-tea.

The springboks graze peacefully far out upon the veldt; the flocks of sheep and goats are all now trending homewards as the sun sinks, the distant tinkle of their bells wafted now and again across the brown-bosomed plain. Lower sinks the great luminary as he prepares his gorgeous vestments for the night; it is a goodly and a striking scene. The pomp and pride of evening now begins; lingeringly the sun departs, charging and suffusing earth, atmosphere, and mountains alike with a ruddy flush and glow of colour, well-nigh inconceivable to any but African beholders. The red light fades, the mountains change from their glorious panoply of crimson and gold to a mellow purple, and as day departs from the karroo, they pass into sombre darkness. Silence, save for the occasional bleating from the kraals or the cry of a distant jackal, now broods upon the landscape, as we turn half reluctantly indoors. The sun has gone, but shortly the African moon, already risen, will appear radiant and serene from behind yonder lowering pile of Witteberg—

> "Here shall the wizard moon ascend
> The heavens in the crimson end,
> Of days declining splendour; here
> The army of the stars appear."

As we light our pipes contentedly after supper, and look out upon the landscape now drenched in an argent flood of the most chastening and bewitching light that ever shone from heaven, we inwardly bless the destiny that has led us hither, to pass among such scenes. Presently we remember that the springboks have to be hunted on the morrow, and betake ourselves to rest.*

---

* Since these lines were written, I have to lament the untimely death of my old and greatly valued friend, Mr. J. B. Evans, of Riet Fontein, J.P. for the Divisions of Graaff Reinet, Aberdeen, Uitenhage, and Willowmore. In him the Cape has lost one of her foremost and most devoted sons, a perfect example of what a colonist should be. Respected and esteemed by all who knew him (and he was very widely known), his bright example will not soon fade from the memory of his fellow colonists, Dutch as well as British. His only fault was that he worked too hard and unceasingly, and his strong frame was undoubtedly worn out before its time, by a too unflagging energy. It is not too much to say that if Cape Colony contained 5,000 British colonists as energetic and shrewd as the late Mr Evans, it would soon become completely transformed.

Chapter XV.

A MORNING AMBUSCADE ON WITTEBERG.

WHILST staying in Naroekas Poort we arranged, amongst other items of sports, to go overnight to the dwelling of Tobias Verwey, our host's Dutch foreman—who lived in a rude hartebeest-huis amid the mountains—sleep there for an hour or two, get up before dawn, and make our way to a nek in the most secluded part of the mountains, where the klipspringers and rheboks were in the habit of passing, soon after sunrise, from their night's drink at a fountain in one of the deep kloofs below. Four of us, therefore, having sent on a Kaffir with our blankets, rifles, and some cartridges, took our shot guns, and in the afternoon wended our way quietly along the mimosa valley, and thence up the kloof that led to Tobias's. It was a glorious day, one of those still afternoons of late spring—warm, yet not too warm, for the air was tempered by a crisp freshness, that later on fades before the glowing heat of summer.

We had calculated on a shot or two as we went along, and were not disappointed. As a spring-haas (jumping hare), that had issued from its shelter somewhat earlier than is usual with these animals, bounded away among the thorn trees to our right, two guns rattled out, and over tumbled the hare. Soon after we noticed, for the second time in our stay, some of the cuckoos that were just beginning to arrive in these regions, known in the Colony by the quaint name Piet-mijn-vrouw—dark grey birds, with reddish breasts and dirty white stomachs with black cross-bars. Le Vaillant calls this bird *Le coucou solitaire*, and Cuvier, *Cuculus solitarius*. Its Dutch name, Piet-mijn-vrouw, is obviously bestowed upon it from its call, which sounds not unlike those words. We had noticed, a day or two before, another and rarer cuckoo—Klaas's cuckoo—Mietjé of the Boers, *Le coucou de Klaas* of the bird-loving Le Vaillant, who named it after his favourite Hottentot. This lovely cuckoo, distinguished by its brilliant green upper plumage, its snow-white under parts and its white tail, centre barred and tipped with green, is not common in the Witteberg valleys, and the specimen we shot was probably the forerunner of the numbers that move to the Knysna Forest towards the end of the year. The green back of this charming bird has a very noticeable coppery metallic sheen upon it, and further, the white lines over each eye, and the green splash of feathers on both sides of the chest, add to its distinction. Le Vaillant's enthusiasm over this little beauty is not to be wondered at. The hen bird is chiefly distinguished from her mate by the brown markings upon her green back.

Not long after we had secured the spring-haas, a pair of purple herons, evidently winging their flight to our river, turned off from their course almost immediately overhead. They were flying carelessly and low, and had not noticed us in the mimosa grove, hereabouts pretty thick, and for one of them the mistake was fatal. We had not killed one of these fine birds since our arrival, and Frank's gun brought down the male bird neatly enough. This bird, the *Ardea purpurea* of Linnæus, is abundant in the Colony; it measures in length very nearly three feet, and in colouring stands pre-eminent among its numerous congeners—for the Cape boasts of eleven true herons (*Ardeæ*), though not all are commonly met with.

It is a delightful country, this Kaap-land as the old Boers called it, if you have leisure enough to explore it; and pastoral farmers out here have, after all, frequently but to sit down quietly and see their flocks increase around them; while to a collector, with plenty of time on his hands, I can imagine nothing more enticing than six months spent in the Eastern province of the Cape Colony, or beyond the Kei in Kaffraria proper (both of them countries not yet half ransacked by naturalists). Presently, as we walk on, a wild cat of some kind jumps out from some bushes in front, and tears away into covert, where it is useless to follow. It was too quick for us, and we lose an opportunity of slaying one of the pests of our much-harassed poultry-yard. But "vorwarts" is the word, and we push on through the deep kloof of which I have previously spoken, and as the sun falls in a warm glow behind the hills, tinting gorgeously the brown rocks around, we reach Tobias's little reed and plaster habitation, where we shall sup and rest for a few hours.

Tobias's vrouw, an excellent but somewhat retiring Dutch lady, of rotund and sturdy figure, is preparing a capital supper for us, and we can already scent from afar, with appetising sensations, that capacious flesh-pot, in which is stewing a *mélange* of francolin and buck, procured by Tobias for our special use and behoof on this occasion. We stroll with the little Dutchman about the premises, note the horses he has in the kraal, hear that a troop of zebras has been seen running on the mountains close to our own far-ranging stud; hear, too, of the untimely demise of a weakly foal killed by a leopard; inspect the patches of oats and melons in the alluvial bottom ground; and then supper is ready.

There are only two compartments, for they cannot be styled rooms, in Tobias's house—in reality it is one apartment, with a rough partition of boards running half-way up to the roofing which shuts off the bedroom. In one corner of the dining-room—if I may so call it—there is an ancient Boer waggon chest, a treasured family heirloom, as these things often are, at least 150 years old, in which is stored most of Tobias's worldly gear, and upon which reposes a huge antique clasp Bible. A rust-bank—a rude sofa of wood, covered with skins—stands against the wall facing the entrance. Upon the wall hangs a solitary picture—an ancient engraving of one of the old-time Dutch governors of the Colony—Governor Ryk Van Tulbagh, if I remember rightly, who flourished towards the middle of the last century. A plain deal table, two chairs, and an empty case or two, complete the furniture. The floor is of the usual South African mixture of ant-hill clay and cow-dung, set perfectly hard and perfectly clean. Tobias and his wife have already supped, and while he sits on his rust-bank and smokes his pipe, she places on the table a mighty tureen steaming hot. The Cape Dutch are not often renowned for the cleanliness or delicacy of their viands, but our worthy entertainers are exceptions. perhaps Tobias's Huguenot descent has transmitted to him traces of the neatness and the culinary propensities of his Gallic ancestors. The plates are clean, the knives and forks strangely bright, and the stew is delicious. Our meal is finished off with honey literally " out of the stony rock," for the caves in the kloofs around furnish a rich and perennial store of magnificent wild honey. Coffee and some very decent Boer brandy and water wash down our repast, and pipes and Boer tobacco—the latter very good of its kind—wind up the feast. Then we have an hour's chat with Tobias, chiefly of anecdotes concerning the fierce Bushmen, who, not so many years since, infested these mountains and harassed the farmer's flocks; and nine o'clock finds us all asleep—Tobias and vrouw behind their partition, ourselves snugly enwrapped in our blankets on the floor. Our sleep was sound while it lasted, but it seemed scarce an hour since we had dropped off when Tobias awakened us at one o'clock.

It was yet dark within doors, and a light had to be procured. Then we took up our rifles, slung on our cartridge bandoliers, strapped up three of our big bush rugs, and issued forth into the chilly morning. The moon was up, and it was not difficult to distinguish near objects; and as the little Dutchman led the way, we followed him pretty comfortably by two and two up a narrow track, that led first up a small kloof, well clothed with stunted olive trees, through which we emerged upon the rough mountain-side. Then followed a tough, and not altogether agreeable scramble over rocks and through low bushes and scrub until the summit was attained. This hill was flat topped, and we waded across its uneven surface, through long sour grass, for a mile or more. Now we were near the nek, trending from a lower hill that sloped two miles away to a long, deep valley, in which lay the fountain I have spoken of. This nek was a sort of ridge, leading from the lower to the higher mountains; it was shut in on either hand by precipitous rocks, which bent inwards and afforded no sort of foothold even for the chamois-like klipspringer. The mountain whereon we stood, which communicated with one end of the nek, led off in various sharp spurs to other heights, and as it was only by this passage that the more secluded hills could be reached from the water (and the water from them) without a long deviation, the antelopes in this vicinity used it pretty regularly in their nightly journey to quench their thirst. The length of the nek was about 250 yards, its width in the broadest part about fifty yards. Tobias had chosen for our ambush two places, the first about forty paces from the entrance out of the lower ground, the other on the opposite side and sixty yards or so nearer the exit to the higher hills.

It was an eerie walk over these mournful, solemn mountains by the pale moonlight, and we were not sorry, at last, to reach our destination. We were soon snugly ensconced with our backs to a rocky wall, and in front of us a screen of bush, which would effectually conceal us from passing game. With the three rugs we had brought with us we enwrapped ourselves as well as possible, for the air was keen and searching. Silence almost complete lay upon the heights around. Occasionally, the weird, harsh cry of a leopard, on his nightly prowl, could be heard in the ravines below us; and now and again the churring of the goat-sucker could be distinguished as the bird paused in its flight, and from some bush or tree sent forth vibrating notes through the still atmosphere. We had now about an hour before daylight, and all, except Tobias, who kept watch, dozed off. It seemed but a second or two, instead of an hour, before we were nudged, and on the alert again. Rubbing our eyes and looking around, we could perceive, as we turned involuntarily to the east (it is curious, by the way, how instinctively the eye seeks the light), the faintest touch of pallor in the dim blue distance over the neighbouring heights. Charlie and Bob now, under Tobias's instructions, take a blanket, creep cautiously behind the bushy screen for seventy yards, and then cross over to the other side, where they are to lie *perdu* till the sport begins. We have noticed, previously, that the light breeze, which just caresses the bushes, blows towards us, thus making matters a point more in our favour. Tobias, who, in matters of venary, is as keen and acute as a Bushman, had previously assured himself on this point, or our plans might have needed alteration at the last moment. We wait impatiently, for pipes are tabooed, and the light comes charily, and as if begrudged, as it seems to our now expectant nerves. But the buck will probably not pass till the sunlight is well over the hills, so we have, perforce, to smother our impatience as we may.

At length the dim eastern pallor spreads and broadens out. Presently thin streaks of light appear, and then a faint saffron tint imperceptibly creeps up.

Anon, a soft rose flush mantles the lower eastern sky. Slowly the glorious day comes, heralded by sweet forerunners of light, painted in every conceivable tint of red and rose, of amber, blue, and green. Now the mountain tops flush as if with pleasure at the coming of the rising king, glorious streamers flaunt and float upwards in the sky, and, finally, all but broad daylight rests upon the earth. But, surrounded as we are by mountains, we cannot yet actually behold the sun.

With the advent of daylight, Nature begins slowly to open her eyes. First is heard the sweet piping notes of the mountain canary—the "berg canarie" of the Boers (*Amadina alario* of Linnæus). As a rule, the Cape is not famous for its songsters, and the radiant colouring of much of its feathered life but ill supplies the deficiency; but this bird and the common Cape canary (*Fringilla canicollis*) are notable exceptions to the rule. The mountain canary, of which I speak, is a neat little fellow, reddish-brown coloured above and as to its tail; the under parts are white, while the head, throat, and neck are black. The sweet-tongued bird, after hopping about a little not far from us, presently shakes off the heaviness of night, and pulling himself together with some pick-me-up discovered among the seeds and berries, flies hither and thither singing merrily between times. It is not long before others of his species join him and add to the melody.

Soon after, as the sun gets warmer and thrusts away the chill night air, a stout rock-thrush and his mate come hopping along in and out of the boulders that lie in such wild confusion around us. The cock bird, as he ought to do, runs first; he is a fine bold fellow when he thinks he is safe and unobserved, as indeed he mostly is up here; and, unseen, we can admire his ruddy brown back and tail, and his dark blue head and throat. Then, in his innocence, he boldly faces us, showing his fine rufous chest and stomach. Presently, after picking up an insect or two, he perks his head round at his lady, who follows closely behind him, and they suddenly stop, and then, spreading their wings, fly upwards to a projecting rock, where the fuller strength of the sunlight can be felt. Immediately the male bird bursts into a bold and pleasing song; his notes are clear and full, and we listen to him with interested ears, for he reminds us of our own feathered songsters at home in England. Another bird flits from rock to rock close to us on our left—a tiny finch, not much larger than a tom-tit, commonly known in the Colony by its Boer name of streep-kopje (little stripe-head). It is a quaint little grey fellow, black striped upon its upper parts, and having its head plainly marked on either side with white and black stripings. Its piping note is not unpleasant, as it searches hither and thither for its morning meal. Suddenly, starting as if from space, comes soaring above us a great black mountain eagle. We know him at once for a berghaan (cock of the mountain) or dassie-vanger (coney-eater). He is evidently watching some object below—probably the antelopes we are waiting for—and doubtless, keen sportsman that he is, he has been the first of all the birds up here to sally forth from his rocky eyrie in search of a good meat breakfast.

At length Tobias nudges me, and instantly our sporting instincts are aroused and acutely alert.

Listening intently, we hear a kind of suppressed grunt borne on the breeze towards us from below. "De klipbok," whispers Tobias, and we cock our rifles quietly. For a few seconds we wait with feverish impatience; still nothing appears.

But now, on a sudden through the sweet morning air, picking its way daintily through the long sour grass and over the rough rocks—its proud little head well erect, the sunlight glinting on its short stiletto-like horns, its moist, round muzzle searching the breeze to catch the faintest breath of suspicion, its large ears anon cocked forward to act as auxiliaries to its keener sense of scent—comes a sturdy klipspringer. How magnificently the little plump fellow moves; how gallant is his port; how bonny his soft brown eyes, and the rich warmth of his olive-brown tinted coat. Just behind him trips the ewe, and near to her again another pair. Close at the flanks of the ewes run a pair of the daintiest little fawns that ever mortal set eyes on. These little creatures skip and spring in sheer gaiety of heart in the most bewitching manner. Their playful leaps and antics are simply astounding. They seem scarcely to touch the rocks that for one fleeting thousandth part of a second give them foothold. Nothing in nature—not even the chamois of Switzerland or the bighorn of America—can excel these rock-lovers in poetry of motion, in grace of attitude, or in the careless daring of their leap; they fly over rather than touch the rocks and earth. A glance at Tobias warns us not to fire. The ewes, with their young at heel, cannot be shot; and it were a shame to slay the gallant rams, whose services may be needed—who shall say how soon?—to protect their families from some prowling wild cat or swooping eagle. Fifty paces behind this charming family group comes a clump of two more rams and three ewes—two of the latter barely full grown. These are marked for our quarry, for they have no young. Motionless as the rocks we lie till the first party is gone, and then, as the last ones trot slowly past, we let fly. They are scarcely thirty-five paces from us, and at the discharge the two rams leap forward—one dead, the other severely wounded; while one ewe, with a broken shoulder-blade, is down. Three seconds later, we hear two reports away on our left, and afterwards we find that, unfortunately, one of the rams of the family party had fallen to the shot. The wounded ram ran on as far as the second ambush, where he was stopped by Bob. It is not often that so many as ten or eleven klipspringers are found together; usually they are seen in pairs, or perhaps as many as four together, but occasionally, as in this instance, when returning to their favourite habitat from water, they may be happened upon in a small "clumpje," as the Boers call it.

But Tobias assured us that, despite the shooting, we shall see yet more game—klipspringer or rhebok—and, true enough, in three minutes there is a helter-skelter, and seven or eight larger antelopes, red and grey coloured, came flying past in a very tempest of fear and fright. They were evidently scared by the noise of the firing, which (as Tobias afterwards told us) rolled and re-echoed round the valley bewilderingly about their ears. There were five of the vaal (grey) rhebok, and three of the rooi (red) kind. These latter usually keep to the lower ground, the vaal rhebok to the higher; but, as in the present instance, when alarmed, they will temporarily unite till the danger is past. Bang! bang! bang! a kill and a half and one clean miss (alas! mine), for the game is up and past like lightning. A rooi rhebok, smitten by Tobias's bullet, is down and struggling; a vaal rhebok, with one of Charlie's bullets clean through him, leaps in the air almost like a springbok, and presses on, hard hit. He, too, is secured at the second ambush, both rifles there, by a mistake, being emptied at him. Thus the rest of these antelopes escape, leaving only two of the troop behind them. We wait five minutes longer, and then we know the shooting is over; no more bucks will now pass. We whistle to our friends in the higher ambush, and gather together the game. As we look across a broad deep kloof to another range, Tobias points to something in motion. It is a troop of nine zebras, which, scared by the firing, are tearing away over the mountain at headlong speed. They are soon lost to view. These splendid creatures, naturally wary to a degree, are, from much persecution, extremely hard to find, and only at early morning are to be seen within some miles of this point. Occasionally, however, they will run with the horses, mules, and donkeys, that range over the hills, and have even been driven down and captured in this way.

While we light our pipes, and Tobias, assisted by two of our number, proceeds to gralloch the slain antelopes, we others examine his rifle. He is not a rich Boer, or he would not have taken service with our host as overseer, and he possesses no newer weapon than an old "roer" that belonged to his father forty years ago. It is a wonderful gun, an ancient smoothbore, with a barrel as long as a yard-arm almost, and it goes off with a report like a cannon. Its spherical bullets, eight to the pound, make a nasty hole in a buck, and their handiwork is plainly apparent.

Six buck—four klipspringers and a brace of rhebok — form the bag, as we think a capital morning's work; for, stalked in the usual way, these mountain dwellers require an infinity of care and trouble to procure; yet, when secured, the reward of your labour is sweet, and a mountain buck is well worth the stalk, fatiguing though it may have been. The only objection to the fun we have been indulging in, is, that it is too quickly over. The rhebok are hidden with stones and bush till a horse can be sent up for them, and each of us shouldering one of the four klipspringers, we make our way down hill towards breakfast. The mountain air up here is bracing and most delicious, and, by the way, how confoundedly hungry it makes one. Here we are nearly 5,000 feet above sea level, and the health-giving breeze seems to send a tingling sense of life and buoyancy hurrying through one's veins. In these regions and amid such surroundings one *lives;* in the dense and sickly atmosphere of cities one exists. Can there be a comparison between the pleasures of such a life and such an existence? I trow not. They who have tasted the pleasures of this clear healthy atmosphere and gloriously beautiful country, will agree with me that six months here are worth long years of dwelling in towns. Yet, on the other hand, I suppose custom can stale even the pleasures of such an existence; at all events, a glance at many a listless, stolid, Dutch Afrikander would seem to imply it.

We were not long in getting down hill to breakfast. On our way we noted a handsome umber-brown lark — "dubbelde leeuwirk" the Boers call it. Le Vaillant knew it, and christened it "l'alouette á gros bec"; its ornithological title is *Alauda crassirostris*. Other larks there are in the Colony, but none of them have the splendid song of an English skylark — more's the pity. A flight of red-wing starlings passed over our heads, flying, doubtless, to some fruit or berry not far off. These birds have dark blue bodies and handsome rufous-tinted wings. The Dutch farmers know the bird as the rooi vlerk-spreo; Le Vaillant discovered it, and christened it "la roupenne." These are shocking fruit stealers, and the wine Boers, during the grape season, find them most troublesome neighbours, and wage determined war on them, guarding their crops with boys and guns. These fellows were, I fancy, on their way to the peach orchards of some Boers away over the mountain. A little lower down, a rock-lizard, about a foot long, flashes across our path. It is a great, gaudy creature, having a bright blue head and back, and a wonderful rose-coloured throat. Its skin is marvellously smooth and soft to the touch. The Boers call it the "blaauw-kop salamander" (blue-headed salamander), and look upon it with feelings of awe and horror. They will tell you solemnly—and they verily believe it—that this reptile is deadly poisonous, and that from it all the snakes obtain and renew their poison.

The lizard, in truth, is perfectly harmless. But

the Cape Dutch—except the better families immediately near Cape Town and its vicinity—have some extraordinary superstitions. For instance, the Boers of our neighbourhood at Naroekas were perfectly convinced that there existed in the mountains, near De Beer Vlei (about ten miles off), a fabulous monster having the head of a rock-rabbit and the body of a snake. After all, considering the amount of superstition yet existing in remote parts of England, the wild beliefs of these Dutch Afrikanders, isolated as they have been from the world for 230 years, are not to be wondered at. We were not long in reaching Tobias's house, where we deposited our loads. Then, taking our shot guns, we had a turn for a couple of hours in the kloofs around at the grey-wing partridges (*Francolinus afer*), securing five-and-a-half brace. After this we attacked and demolished, soon after seven o'clock a.m., an excellent breakfast, and then, with Tobias, spent an hour or two getting honey from a deposit in the rocks near at hand. In the afternoon, we returned home contentedly with horses well laden with venison and honey.

CHAPTER XVI.
THE PRESENT DISTRIBUTION OF THE ANTELOPES AND LARGER GAME OF CAPE COLONY.

WHEN Governor Jan Van Riebeek set foot on the shores of Table Bay in 1652, hoisted the Dutch flag, and took possession of the soil, he found a land literally teeming with game, and absolutely virgin, so far as Europeans were concerned, for the hunter. Before him stretched precipitous mountains, wide karroos, and rolling uplands, crowded, as no other country was ever crowded, with the noblest, the most beautiful, and the most stupendous game that nature in her bounteous moods ever produced. The elephant, the rhinoceros, the buffalo, the quagga and zebra, the lion, the leopard, and the ostrich, roamed in profusion and in undisturbed freedom over the country to the very margin of Table Bay—nay, the strand wolf (*Hyæna brunnea*) even found its food from the dead whales and other fish cast upon the sea-shore; while in every river the hippopotami wallowed in countless plenty. As for antelopes, their numbers were literally as the sands of the sea. Within the old and more commonly known limits of the Cape Colony (that is, taking the Kei River as the eastern boundary, and the Orange River as the northern), with perhaps half-a-dozen exceptions, every variety of antelope to be found between the

Zambesi and the Indian Ocean there had its habitat. The magnificent eland, an animal surpassing in weight and stature the heaviest ox, and yet possessing all the truest points of the antelope tribe, must have been found in extraordinary profusion, for its name to this day lingers upon river, mountain, plain, and kopje in every part of the Colony.

The rare roan antelope, now approaching extinction in South Africa altogether, restricted as it ever is in its habitat, was common in Swellendam, where it survived till the end of the last century; the gemsbok or oryx abounded on every open karroo, the curiously-pied bontebok was found in vast numbers; and the black wildebeeste, blessbok, springbok, and hartebeeste upon the open plains literally covered the face of the earth. Indeed, but for the brandtsickte or burning disease which ravaged their herds, they must have become far too numerous for the country to support. All the smaller antelopes were found in unexampled plenty. The Hottentots and Bushmen, who at that time held the territory now known as the Cape Colony, though they obtained their food supplies from the game around them, could from the nature of their weapons (the bow and assegai only), if left to their own devices, never have made much impression upon the vast natural game preserves of their land.

But since the Dutch landed and firearms were introduced, the history of the fauna of Cape Colony, and, indeed, of South Africa generally, has been one continual record of ceaseless, wanton, and shameful slaughter. It will be said that our own countrymen, as well as the Dutch, have had a great hand in this slaughter. To this I may point out that those of our own race (of course with some exceptions) have more sportsmanlike ideas than other nations, and have not joined in the useless slaughter and skin-hunting forays that the Boers have invariably and incessantly indulged in.

Perhaps there is no better illustration of Boer wastefulness and of wicked destruction of animal life than the rolling plains of the Orange Free State. Thirty years ago, these plains literally swarmed with Burchell's zebras, quaggas, wildebeestes (gnus), blessbok, and springbok. The Dutch found that the skins of these animals brought them temporary wealth, and in consequence scarcely a head of game can now be found in that country; the bones of these beautiful creatures lie literally whitening the veldt—all have vanished, and the life of the Free State Boer is now robbed of half its former charm. In the Cape Colony, English farmers have, to my own knowledge, in many places succeeded in preserving some of the rarer game now left; and in the case of the koodoo, zebra, and other animals, have actually increased their numbers, as they certainly have prevented their extinction. In the district of Uitenhage, there are certain contiguous farms of Englishmen where the koodoo is preserved, and hunted only a few days in the winter of each year. As a consequence, it is now fairly plentiful; and it is a curious fact that these koodoos never wander upon the farms of the neighbouring Dutchmen, where their extinction would be quickly and surely wrought.

It may be doubted exceedingly whether the Dutch, who first landed at the Cape and became acquainted with the many species of antelopes then abounding there, were sportsmen to the manner born. Rather would it appear that two hundred years of the chase, on the most glorious hunting grounds ever provided by nature, have made them the fine shots and ardent hunters they now are. Certainly, if we may judge from the absurd and anomalous nomenclature bestowed by these early settlers on the game of South Africa, they must have been the very cockneys of their period.

For instance, the word eland signifies an elk, but the difference between this, the largest of the antelopes, with its stout straight horns, and the true elk with its heavy branching antlers, could not well be wider. Every antelope almost seems to have been called a "bok," which, translated, literally means a goat, and from this misnomer many absurd mistakes have been made by the uninitiated. Thus, Dr. Brookes, who appears to have gleaned most of his information from Kolben, who visited the Cape in 1705, in his Natural History, published in 1763, gravely classes many of the South African antelopes among the goats, as the blue goat (the koodoo, then called the blaauwbok), spotted goat (bontebok), grey goat (grysbok), diving goat (duykerbok), etc.

The name gemsbok, literally translated, means chamois goat, and is perhaps the worst specimen of Boer misnomer. The gemsbok or oryx inhabits waterless and open karroos, is about the size of an ass, and has long straight sharp horns, more than three feet in length; while the chamois, not found in Africa, lives in the mountains, is of no great size, and has short horns turning over at the points. How the early Dutch could have traced a similarity between these animals, or what their ideas of the European chamois could have been, passes understanding. The koodoo retains its native name; but the hartebeeste signifies hart or stag-ox, presumably from its non-bovine characteristics. Wildebeeste (the gnu) means wild ox, not a happy simile; the blessbok takes its name from the white "bless" or blaze on its face.

The tsesseby, or sassaby, is called by the Dutch the bastard hartebeeste; the sable antelope, "zwart wit pens"—*i.e.*, black with white belly; the roan antelope, blaauwbok (blue bok), sometimes bastard gemsbok, or bastard eland, both singularly inappropriate names; while among the larger game, kameel (camel) standing for the giraffe, tiger for leopard, and zee koe (sea cow) for hippopotamus, are all somewhat uncouth and far-fetched. Evidently, the Dutch pioneers did not carry a practical naturalist with them in their wanderings. And yet, once in South Africa, all these strange and uncouth names seem to come naturally enough to the lips, and indeed, I think by most hunters (English as well as Dutch) are regarded with a sort of affection; for they recall vividly many a happy hunting ground, many a gallant head of game, and many a scene of natural beauty, such as in its own wild picturesqueness, perhaps no other other country can surpass. It will, perhaps, better serve the purpose of this chapter if I enumerate the large game and antelopes that found a home in the Cape Colony one hundred years ago, and then endeavour to trace slightly the decline and disappearance of such as are now extinct within its limits, and the present occurrence of such as remain. For this purpose I will take the colonial boundary as it was best known—viz., having the Orange River for its limit to the north, and the Kei River to the east, and not including Griqualand West and Transkeian Kaffraria, which are now actually under Cape Colonial Government.

One hundred years ago, then, there were to be found within the territory I have indicated the elephant, black rhinoceros, hippopotamus, buffalo, zebra, quagga, lion, and leopard; and of the antelopes, the roan antelope, eland, hartebeeste, koodoo, gemsbok or oryx, black wildebeeste or gnu, bontebok, blessbok, springbok, rietbok, vaal or grey rhebok, rooi or red rhebok, klipspringer, duyker, boschbok, grysbok or grys steinbok, steinbok, oribi, and the blaauwbok or kleenebok. From this category I have omitted the white rhinoceros (rhinoceros simus) and the giraffe. Both these animals are stated to have been found in the Colony at a remote period; and, indeed, Barrow, in his travels, asserts that the former of these animals was plentiful in Great Bushmanland in 1796. From the nature of the country, it seems not improbable that this may formerly have been the case, unless scarcity of water in that parched region interfered; but, as the point seems involved in doubt, I have refrained from including this animal. As regards the giraffe, there is no certain and reliable evidence as to its former existence within the Colony; but it may be pointed out that this animal has, within the last one hundred years or so, been frequently shot within a day or two's journey north of the Orange River (see Paterson's, Le Vaillant's, and Campbell's travels in 1777, 1784, and 1813 respectively); and, sharply and curiously defined as are the geographical limits of occurrence of many African animals, there seems no sound reason, other than nature's caprice, why the camelopard should never have crossed into the colonial limits. The rude drawings in the Bushman caves near Graaff Reinet and in other parts of the Eastern province, and an old tradition of the Hottentots, that this animal was formerly found in the Amaebi or thorn country (now part of the Queenstown division), are, at all events, some slight evidence in favour of the theory that the giraffe in ages past browsed south of the Orange River.

The Elephant (*Elephas Africanus*) is happily still to be found in Cape Colony. In the Knysna Forest, where they are preserved, and where the Duke of Edinburgh shot a fine bull in 1870, they are still not uncommon. In the dense bush-veldt thickets of the Eastern province, especially between Uitenhage and King William's Town, where they are almost inaccessible, and in the Zitzikamma Forest and Addo Bush, they manage to maintain their ground. Even down to 1830, the elephants in this part of the country were vigorously hunted for their ivory, and must have been then numerous; and Barrow, in 1796, mentions that between Bushman and Kareeka Rivers the country was a very nursery of elephants, and that in this region one Rensburg saw a troop of four or five hundred crossing a plain between the bush veldt.

The Black Rhinoceros (*Rhinoceros bicornis*).—A hundred years ago this rhinoceros must have been common in the neighbourhood of the Great Fish River, and between it and the Kei; and in the rough bushy country west of the former river, not far from its mouth, Barrow speaks of it as being abundant. The last specimen seems, according to Gordon Cumming, to have been seen as late as 1849 in the Addo Bush. In the latter half of the seventeenth century, and during a greater portion of the eighteenth, this rhinoceros roamed freely over the whole of Cape Colony, where the pasture suited; and its Dutch name rhenoster yet remains on many a hill, river and fountain.

In 1851, there lived on the Great Fish River an old Boer, named Bezuidenhout, who in his youth had killed many rhinoceroses on the eastern borders of the Colony, and in Kaffraria itself. The old man at that time was about eighty years of age, and still affectionately retained, long after his more civilised fellow-Dutchmen in the Colony had discarded them, his immense long "roer" (elephant gun) and "veldt broeks" (literally, field breeches) —the ancient garments of leather worn by the old-world Dutchmen of the Cape. Many a good hunting story could the old man tell, and amongst them was one in which the "veldt broeks" had played an important part. One day, when out shooting, Bezuidenhout was charged by a wounded rhinoceros, which caught him with its horn between the legs. The Boer, miraculously, was unhurt, and managed to cling to the animal's head, while it rushed madly onwards for nearly a mile, as the old man always stoutly swore; at length, in entering a grove of stunted trees, the frightened rider managed to clutch a branch, and hoist himself out of harm's way. The old man always attributed his wonderful escape (for he suffered only a few bruises) to the strength and thickness of his leather breeches. The story seems almost incredible, but those who knew the old man and who told it to me believed in it; and amongst the marvellous escapes of African hunters, it is not altogether improbable. Old "Veldtbroeks" (as he was called) and his rhinoceros story were well known in the Eastern province five-and-thirty years ago.

The Hippopotamus (*Hippopotamus amphibius*), although I believe it only disappeared finally from the Great Fish River within the last twenty years, is now only to be found in the waters of the Orange River, and there only to the westward of the Great Falls. Formerly it abounded in every river in the Colony. I believe a few are still to be found in the rivers of Kaffraria.

The Buffalo (*Bubalus caffer*) is still fairly numerous in the denser parts of the bush veldt of the Eastern province. In the Knysna Forest, the Zitzikamma Forest, on the slopes of the Great Winterhoek range, and between Sunday River and the Great Fish River, and the Great Fish and the Kei Rivers, these fine animals find sanctuary. A few are shot every year, especially between Wolve Kraal and Uitenhage; but the thickets in which they lurk are so thorny and impenetrable, that hunters find their sport attended with much pain and peril, from the lacerations of the bush and the dangerous nature of the buffaloes in these gloomy thickets.

The Zebra (*Equus montanus*), as I have pointed out in a former chapter, is still fairly abundant on some of the mountains of the Colony, notably the Zwartberg, Witteberg, Great Winterhoek, Baviaans Kloof, and Sneeuwberg, and near Cradock and a few other places.

The Quagga (*Equus quagga*), or quacha, is now quite extinct within the Colony. Formerly it abounded on every plain, and with a very little preservation, might even now be adorning the landscapes. The last observed on the Great Karroo were three that still remained, in 1858, near the Tiger Berg, in the neighbourhood of Aberdeen. In a subsequent chapter I have dealt at some length with the decline and fall of this handsome and interesting quadruped. I was told not long since that quaggas were still to be found in the Outeniqua district, but I subsequently ascertained that my informant had confounded them with zebras, which did undoubtedly exist there. It is a moot point whether Burchell's zebra (*Equus Burchellii*) was not formerly found within colonial limits. Paterson's "Journeys," 1777-8-9, would certainly seem to imply that such was the case; while on pages 318 and 319 of his "Travels," Barrow speaks of an animal of the zebra kind, yet differing from the quacha and true zebra, seen near Bambosberg at the end of the last century by several Dutch hunters, which in several respects answers to the description of Burchell's zebra. Naturalists have, however, laid down, chiefly on the authority of Cornwallis Harris, that this species never came south of the Orange River. It is much to be regretted that no record of South African travel exists between the time of Peter Kolben, 1705, and Sparrman, 1772. Such a record would, probably, have done much to establish the true geographical distribution (now for ever in doubt) of several animals, notably the giraffe, white rhinoceros, and Burchell's zebra. In 1705, Kolben's time, the present colonial territory and its fauna, except within less than one hundred miles of Cape Town, were very little known.

The Lion (*Felis leo*).—Since January 23rd, 1653—when according to the quaint old Dutch journals, now preserved in the Colonial archives, "This night it appeared as if the lions would take the fort (*i.e.*, the present Cape Town) by storm"—this animal has been gradually, but surely, exterminated or driven back. In the same year, 1653, Governor Van Riebeek encountered a lion in his garden; and the king of beasts long held his ground against the Dutch intruders. Even between 1825 and 1830, when Steedman travelled in the Colony, the lion was exceedingly plentiful; but probably from the date of the introduction of percussion caps its downfall proceeded more hurriedly. It would seem that the lion had finally disappeared from the Colony by 1850 or thereabouts, or perhaps, in the remote parts of Bushmanland, a little later. The present Queenstown district was one of its last strongholds.

The Leopard (*Felis pardus*) is still common in nearly all the mountains of the Colony, and is, from its habits and habitat, principally kept down by means of poisoned meat. It is, however, occasionally shot, and occasionally severe and even fatal accidents happen in encounters with these dangerous brutes.

The Roan Antelope (*Hippotragus leucophæus*).— This magnificent and exceedingly scarce antelope— the bastard eland, or bastard gemsbok of the Dutch, sometimes also called by them in bygone days the blaauwbok—was formerly found within the Colony, but apparently only in the Swellendam division, and in the neighbourhood of the Breede River.

According to Barrow, the last of this species was seen about 1786 ; but other authorities (Steedman amongst the number) place its final disappearance in 1799. Between these dates probably lies the true period.

The Eland (*Oreas canna*).—Formerly one of the most abundant of all the antelopes, this, the largest and in some respects the noblest of the antelope kind, has long disappeared. From the excellence of its venison and the ease with which it could be pursued and slain, it fell a speedy prey to all hunters. It is nothing short of a disgrace that an animal so easily preserved, even in captivity, should not have been protected in some way by the colonists. Even now a few elands might, with a very little trouble, be re-introduced, and in a few years would quickly multiply and prosper in their ancient habitats. In Campbell's time (1813) this antelope was common in the north of the Colony, and was repeatedly seen chased by three or four lions. It probably lingered till between 1840 and 1850 in the waterless deserts of Bushmanland. Barrow found them in abundance on the Great Karroo and other plains, and shot a specimen six feet six inches in height.

The Hartebeest (*Alcelaphus caama*).—This antelope has almost disappeared entirely from the Colony, but, as I hear, a few are yet to be found along the Orange River between Little Namaqualand and the Orange River Falls. Seven years ago a few were shot in that region. It seems to have been fairly abundant till 1840 or thereabouts. From its swiftness and endurance, it is difficult to understand why it has not lingered longer.

The Koodoo (*Strepsiceros kudu*).—It is a pleasure

to state that this noble antelope, one of the most beautiful and striking as well as one of the largest of its race, is still fairly abundant in the Eastern province. Of late years, indeed, owing to the preservation of British farmers, it has even largely increased in numbers. It is found on the slopes of the Great Winterhoek Mountains, and on the Messrs. Hayward's farms in the district of Uitenhage, thanks to their fostering care, fifteen or twenty may now be seen in a day's hunting. It may be noted, however, that these koodoos are only hunted for a short time in each winter. It is hardly necessary to mention that, if left to the tender mercies of the Boers, this antelope would soon be mercilessly shot off. Last winter a bull koodoo, with horns four feet long, and weighing nearly 500 lbs., was shot in the Uitenhage district. They are also found in the bush-veldt near the lower portions of Gamtoos River, Sunday River, and Fish River, as well as in the neighbourhood of the Zitzikamma Forest, and at the eastern end of the Zwartberg range.

The Gemsbok or Oryx (*Oryx capensis*).—This rare and singular antelope—undoubtedly the prototype of the fabled unicorn, which it resembles in nearly every particular, except that its long horns (which, by the bye, seen in profile, look like a single horn) are set back instead of forward—is very nearly extinct in Cape Colony. Seven or eight years since, two of the last were shot in the very north of Great Bushmanland (now known as Calvinia), towards the Orange River, by members of a Government Survey Expedition, and a few still linger there and on the eastern confines of Little Namaqualand. Formerly, the gemsbok was plentiful on every karroo of the

Colony. Even between 1825 and 1830 Steedman found them in fair numbers on the Great Karroo: while in 1843 Gordon Cumming hunted them on the plains west of Colesberg, and writes in ecstatic terms of the beauties of this striking and interesting antelope. The oryx, in nearly the same form as in South Africa, is found in Northern Africa, and it would seem that the idea of the unicorn was first taken from the representations of this animal found on Egyptian and Persian monuments.

The Black Wildebeest or White-tailed Gnu (*Catoblepas gnu*).—Formerly found in vast numbers on the plains, this antelope now only remains upon one or two farms in the north-west of the Colony (Victoria West), where it is carefully preserved. It was still fairly abundant on the northern plains, between 1850 and 1857, especially between Colesberg and Hanover. Unless the few now preserved are well looked after, and suffered to perpetuate themselves, the Cape of Good Hope Government is in danger of losing, at one and the same time, both the ancient supporters of its coat of arms—the gemsbok and black wildebeest.

The Blue Wildebeest, or Brindled Gnu (*Catoblepas gorgon*), blaauw wildebeest of the Dutch, has been hitherto erroneously omitted from catalogues of the large game formerly to be found within the Cape Colony. It is laid down by Dr. Smith and Cornwallis Harris, in their treatises on the game of South Africa, that this animal was never to be found south of the Orange River, and this theory seems never to have been controverted by naturalists since 1837, when Harris made his expedition from Graaff Reinet to the Tropic of Capricorn. That the blue wildebeest was, however, formerly a denizen, albeit a rare one, south of the Orange River, and that Dr. Smith and Cornwallis Harris were, in this particular, mistaken or misinformed, is proved by no less an authority than Gordon Cumming. In December, 1843, when that great hunter was shooting with Mr. Paterson, an officer of the 91st Regiment, in the karroo country west of Colesberg, in what is now the Hope Town division of the Cape Colony, the brindled gnu is twice mentioned as having been shot.

In his well-known work, "The Lion Hunter of South Africa," Gordon Cumming says: "I despatched one of my waggons to bring home the oryx, and it returned about twelve o'clock that night, carrying the skin of my gemsbok, and also a magnificent old blue wildebeest (the brindled gnu), which the Hottentots had obtained in an extraordinary manner. He was found with one of his forelegs caught over his horn, so that he could not run, when they hamstrung him and cut his throat; he had probably managed to get himself into this awkward attitude while fighting with some of his fellows." And again, a few lines later, he writes: "I lent him (Paterson) Cobus (a Hottentot), and on this occasion his perseverance was rewarded by a noble gemsbok, which he rode down and slew, and also a fine bull blue wildebeest, *which last animal is rather rare in these parts.*" I have italicised these last words. The reputation of Gordon Cumming is, I think, sufficient to fully establish the fact that the brindled gnu was formerly found within the Colony; but that, unlike its *confrère*, the black wildebeest, its occurrence there was rare. It is singular, to say the least of it, that this fact has not (as I believe) been noted and set right by naturalists. It may be added that the blue wildebeest, admittedly rare even in Gordon Cumming's time, south of the Orange River, has, for many years, been extinct in those regions.

Gordon Cumming found the brindled gnu in the territories now known as Griqualand West, the Orange Free State, and Transvaal. The last-named state was, undoubtedly, its chiefest and most favourite *habitat;* but there the Boers have, of late years, played sad havoc with this singular antelope, not so long ago found in countless numbers.

The Bontebok (*Alcelaphus pygargus*). — It is difficult to understand why this antelope, described by old writers as being almost as plentiful as the springbok, should have been one of the earliest to disappear. One of its last resorts were the Bontebok Flats, to the north of the present Queenstown district. Mr. J. B. Evans, late of Riet Fontein, on whose farms I formerly shot in the Colony, told me that in 1851 he knew of seventeen or eighteen being in that locality. At the present day there are a few preserved on private farms in Bredasdorp, Swellendam, near Cape Agulhas, where they have been ever since 1830. Barrow speaks of them as abounding in his day (1796) near the Zee Koe River, between the present town of Hanover and the Orange River, and he further mentions that they had been as plentiful in Swellendam as springboks in the Sneeuwberg district, but were then only seen there in troops of ten or twelve.

The Blessbok (*Alcelaphus albifrons*) is now extinct

in the Colony. It would seem never to have been abundant, except on parts of the northern plains. Steedman, in 1830, speaks of it as being scarce even in that region. Its more natural habitat appears to have been the Orange Free State, whence it descended periodically across the Orange River into the Old Colony. These antelopes do exceedingly well in captivity. I came home from the Cape when there were two on board the ship, and both of them fed upon hay, throve well, and remained in fine condition during the whole of the voyage.

The Springbok (*Gazella euchore*).—I have indicated the present occurrence and numbers of this antelope in a former chapter. Despite the warfare that has been waged upon it, it is still found in considerable plenty on the Great Karroo, and on the plains of the country formerly known as Great Bushmanland, it abounds in large herds. In former days these beautiful creatures were literally innumerable. Barrow says, at page 118 of his travels: " The springbok is met with on the plains of Camdeboo (near Graaff Reinet) in numbers that are almost incredible. A thoroughbred sportsman will kill from twenty to thirty every time he goes out. This, however, the farmer (Dutch) does by a kind of poaching. He lies concealed among the thickets near the springs or pools of water to which the whole herd towards the close of the day repair to quench their thirst, and by firing amongst them his enormous piece loaded with several bullets, he brings down three or four at a shot." A few years since, I still found the springbok in considerable numbers in crossing these very Camdeboo Plains, and a little farther south in thousands; but the inclosure of farms is just beginning in many parts of the Great Karroo, and will have some influence on the free range of the antelopes on this great sun-dried plateau. As an instance, however, of inclosure in some cases tending to preserve game in the Colony, the *Graaff Reinet Advertiser*, of November, 1886, mentions that Shirlands, the property of Mr. John Priest, of that district, was, twelve or thirteen years ago, a piece of waste land abandoned to squatters. Now there are 16,000 morgen (more than 32,000 acres) fenced with wire. Within this fence there are now fully a thousand springboks, where formerly only a few remained "harassed and hunted to death by impoverished, lazy squatters."

The northern parts of the colonial divisions of Little Namaqualand, and of Calvinia, Fraserburg, Carnarvon, and Victoria West, formerly called Great Bushmanland, will, from their arid nature, long provide comparatively undisturbed sanctuary for large numbers of "the showy buck," as the Boers sometimes call this beautiful creature.

The Rietbok (*Eleotragus arundinaceus*) is, I believe, now nearly extinct within the Colony. As far back as 1830 it seems to have been somewhat scarce even in the Eastern province, its natural habitat, and from its nature and habits was easily shot. It is possible a few specimens may still linger in the denser parts of the Fish River thicket, but it is, I think, improbable.

The Vaal or Grey Rhebok (*Pelea capreola*) is, as I pointed out in a former chapter, plentiful on most of the mountain ranges of the Colony, and is likely to remain so.

The Rooi or Red Rhebok (*Eleotragus reduncus*) is also fairly plentiful, especially in the mountains of the Eastern and Midland provinces. It resembles the vaal rhebok almost exactly, save in its colour, which is of a bright red brown. It chooses for its haunts the lower parts of the mountains, where the "rooi" grass is abundant, and where its own colour harmonises almost exactly with its pasture. The vaal rhebok, as I have mentioned, is found higher upon the mountains, where the "zuur veldt" is of a greyish colour, assimilating with its own coat. There is a very curious circumstance in connection with these two antelopes, which I think has not been previously remarked by naturalists.

In some localities, where the "zuur veldt" clothes the upper parts of the mountains, and the "rooi" grass the lower portions, the vaal and rooi rhebok may be found on the same mountain-side, but each adhering rigidly to its own peculiar pasturage. When the hunters come upon the ground to shoot, the rooi rhebok immediately fly from their lower slopes to the higher ground of their grey brethren, and the two species are seen galloping in close company over the mountain heights. If the hunter rests quietly after his shot and looks about him, he will presently see the two kinds of antelope, as soon as they think they may safely do so, separating, the rooi rhebok quitting the "vaal" pastures, and betaking themselves again to their own feeding grounds. To this habit they invariably adhere, and will not delay their departure an instant longer than their safety admits of. If the vaal rhebok in turn are driven out of their own ground, they pursue exactly the same tactics, and will on no account remain for long in their red brethren's territory. It is a remarkable fact that the fawns of none of the smaller antelopes, with the exception of the springbok, can be reared without falling victims to a disease of the eyes, which invariably ends in blindness. Experiments have been made with scores of the young of the steinbok, duyker, and rhebok, but always with the same result. It having been suggested that, as the natural haunts of these animals are of a bushy or partly bushy nature, the fawns suffered from too much exposure to the sunlight, the experiment of placing them in dark sheds and other shelter was tried, but the blindness ensued in these cases in just the same way. At the time I stayed with Mr. Evans he had a beautiful young fawn of the vaal rhebok, which was perfectly tame and almost lived in the house. At that time the ophthalmia was just beginning to show itself, and I heard afterwards that the pretty, gentle creature became quite blind, and had to be destroyed.

With the young of the springbok, on the contrary, no trouble whatever is experienced; they are never afflicted with any affection of the eyes, and invariably grow up strong and hardy. It would seem from this circumstance that the question of natural cover and bushy habitat has, after all, something to do with the ophthalmia in the other antelopes, and that the springbok, loving as it does the open plains, and exposed as it is to the full glare of the sun, bears with impunity the changed surroundings. Unquestionably, too, I think the springbok is the hardiest of all the antelopes.

Perhaps my readers may have some explanations to offer upon this subject.

The Klipspringer (*Oreotragus saltatrix*), as I have mentioned in other chapters, is found in plenty on the most precipitous and rugged mountains in all parts of the Colony. It may be noted that the young of this antelope are the most difficult and troublesome to rear of all the South African family —in fact, they can hardly be kept alive in captivity. I suppose the extreme solitude and freedom of this little beauty in its natural state has something to do with the fact.

The Duyker or Duykerbok (*Cephalopus mergens*) is common almost everywhere in the Cape Colony, in deep bushy bottoms and kloofs, or in thick bush.

The Boschbok or Bush Buck (*Tragelaphus sylvaticus*) is abundant in the broad fringe of thick bush-veldt extending along the southern and south-eastern coast line to Kaffraria. It affords excellent sport to the colonists, and at the annual Easter hunt near Port Elizabeth, large numbers of this antelope are shot. It is one of the most plucky and determined of its kind.

The Grysbok or Grys Steinbok (*Nanotragus melanotis*).—This variety of the steinbok is found plentifully in the same localities as the boschbok. Very occasionally it is encountered near the vleys sometimes found on the tops of mountains higher up the country, but very rarely; and it may be doubted whether these are not stray antelopes that have wandered or been driven out of the bush-veldt fringe where they are usually found.

The Steinbok (*Nanotragus tragulus*).—This beautiful little creature is to be found freely all over

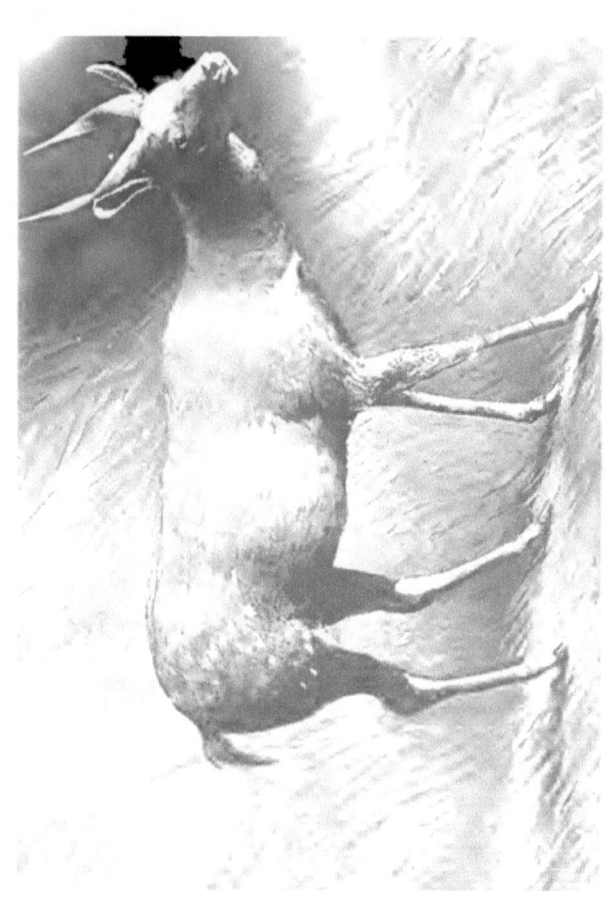

the Colony where slight bush or kopjes (stony bluffs), sparsely covered with bush, are plentiful.

The Oribi or Bleekbok (*Nanotragus scoparius*) was formerly abundant on the grassy plains of the north-east portion of the Colony. It is now comparatively rare, but may yet be found in the divisions of Somerset and Bedford, and one or two other localities on the Eastern frontier.

The Blaauwbok or Kleenebok (blue buck or little buck) (*Cephalopus pygmæus*), the tiniest of all antelopes, hardly bigger than a guinea-pig, is found plentifully in the bush-veldt line of the eastern seaboard, in the same localities as the boschbok and grysbok.

In addition to the game I have above enumerated, the bosch vark (bush pig) or wild boar; the tiger wolf of the colonists (*Hyæna crocuta*); the strand wolf (*Hynæa brunnea*); and the aard (earth) wolf (*Proteles cristatus*); as well as the curious Cape hunting dog, the wilde honde of the Dutch (*Lycaon pictus*), are still found within the Colony. The hyænas are, however, happily for the farmers, who exterminate them unmercifully, becoming scarce; while the wilde honde is now principally found towards the coast line of the Eastern province. Jackals are common everywhere. Here and there on the more remote karroos, the wild ostrich may still be found. With this slight and imperfect sketch of the principal fauna now remaining to the Old Colony, I will conclude this chapter.

I have been induced to take up this interesting subject, for the reason that it is long since any approach towards settling the present distribution of these animals, under the changed conditions of colonial life, has been made, and that the wildest ideas are often prevalent as to what game is or is not to be found within the colonial marches. There is even now in the vast area of the Cape Colony abundance of room for farmers as well as game; and such is likely to be the case for many a year to come. It is devoutly to be hoped that some effort may yet be made towards re-introducing some of the lost antelopes of the Colony, such as the eland, blessbok, hartebeest, and others. With very little protection and care, these animals would soon become plentiful, and afford good sport in their ancient hunting grounds. That long years may elapse before the mighty elephant, the grim and gloomy buffalo, and the stately koodoo are driven from the sanctuaries they now possess in the dense thickets and dark-green impenetrable bush-veldt of the Eastern province—before the last hippo bathes his unwieldy limbs in the waters of the Orange River—or before the flying springbok is finally improved from the face of the Great Karroo, and even from the wastes of Great Bushmanland—is an aspiration that I am sure all sportsmen and lovers of nature will heartily share with me.

The destruction of game in all parts of the world has of late years proceeded at so alarming a rate, that it is not surprising to hear simultaneously of severe and stringent game laws from more than one of our Colonies. The Ontario Legislature, for instance, has enacted that the deer-killing season shall be confined to a period between October 15th and November 20th. No one person is allowed to kill more than five deer in a year, no two persons of the same party more than eight, no three persons more than twelve. Moose shooting is absolutely prohibited for seven years, and, further, no person may kill deer at all unless he has resided in the province for three months and holds a licence. In the United States, owing to the frightful waste of the fauna, strict game laws have been in vogue for some time; moreover, the Americans—Republicans though they are—see that their enactments are rigidly enforced.

Similarly in Cape Colony more stringent laws, directed to the preservation of game, have been recently passed. Within this last year (1888), the close time fixed by the proclamation of 1882 has been greatly extended. All antelopes are now protected in every part of the Colony by a close season, varying from six to seven months—July to January, and August to March—in different districts.

In the division of Caledon, no antelopes may be shot for two years (from the beginning of 1888); in Bredasdorp, no rhebok or klipspringer for three years; and so on in different localities in proportion as the particular game has become scarce.

In Calvinia, formerly known as Great Bushmanland, the rare gemsbok (*Oryx capensis*) and the curious hartebeest (*Alcelaphus caama*), both well nigh extinct within the Colony, are protected from slaughter for three years; while the singular bontebok or pied antelope (*Alcelaphus pygargus*) is similarly protected.

The Cape colonists are wise in time. The game in their territory, even now, through the common sense of the settlers, more abundant than in the Orange Free State, will, by these laws, be greatly increased, and they will be able to look in future with some complacency from their own borders to the Transvaal and Orange Free State, where the skin-hunting Boers have almost completely swept a magnificent and redundant fauna from the face of the land.

CHAPTER XVII.

THE GAME BIRDS OF CAPE COLONY.

MOST people have some vague and shadowy idea of the wealth of South Africa in the matter of the larger game; few in this country probably have any conception of the vast store of the smaller mammals and birds to be found, even at the present day, within the comparatively accessible regions of the Cape Colony. Since the days of Cornwallis Harris and Gordon Cumming, "big game" has been a sort of traditional attribute of the country; but the big game are year by year retreating to the dim and distant regions of the far interior with much more rapidity, probably, than the general public are at all aware of. I have endeavoured to point out in previous chapters that many of the smaller antelopes, in which the Cape Colony yet abounds, are but little known to Europeans, and, for various reasons, unrepresented in any of the zoological collections in England or the Continent. I fear that the game birds, with which the Old Colony is so plentifully endowed, languish in a still more melancholy obscurity. And yet this group of birds, numbering at the lowest computation some nineteen or twenty different species—not counting wild duck, wild geese, rails, and others—can boast of some of the noblest and most striking looking feathered fowl to be found in any country of the world. Within the broad territories bounded by the Orange and the Kei

Rivers, francolins, guinea-fowl, sand-grouse, quails, and bustards flourish in extraordinary plenty, and among the bustards stands pre-eminent the great Kori bustard—the gom paauw of the Dutch colonists—the *Otis kori* of Dr. Burchell, one of the earliest and most scientific of the old Cape travellers. This noble bird, ranging as it does in weight from twenty-five to forty pounds and even more, is of itself sufficient, whether from its merits as a table bird, or from its sport-giving qualities, to provide a reputation for an entire Colony. But this, somehow or other, is not the case, and the game birds of the Cape are, even among the colonists themselves, by no means so sought after as their sporting merits would lead one to expect. It is certain that in this country the whole group are almost unknown, save to the naturalist. Chief among the reasons for this apparent neglect is, I imagine, the fact that hitherto the antelopes and larger game have to a great extent dwarfed and set aside, in the average sportsman's view, this exceptionally large and varied class of birds.

The Boer has never taken any notice whatever of the splendid table birds that have their habitat within his borders, excepting always the kori bustard, and perhaps an occasional koorhaan (another of the bustards), which he can secure with a bullet from his rifle now and again. But the Boer will some day in the distant future awaken to the discovery that antelopes are not so plentiful as they once were; and he may then deign to turn to the francolins and bustards that lie ready to his hand. In the meantime, it is perhaps just as well for the British settlers that the Cape Dutch have not taken thought for their game birds, or, as has happened with much of the nobler game of the country, their extinction might by this time have been partially wrought in certain districts.

It would be interesting to fling aside the curtain that separates the present from the future, and to be able to look forward fifty years, in order to ascertain what manner of man the Boer of that period will be. Not improbably, having shot off the last of his "groote wilde" (big game), he may be found to be bestowing as much care and attention on his feathered game as any English squire; to have devised strict game laws and a stern code of shooting etiquette; and to be pursuing his day's shooting over a brace of highly-bred dogs with as much zest and punctiliousness as an English country gentleman of thirty years back. At all events, let us hope such may be the case. The sportsmen of these islands of ours, who wander nowadays so far afield in search of game, whether furred or feathered, can have but little conception of the diversity of feathered game that lie everywhere at hand in South Africa, or I imagine the Cape Colony would be much more exploited by fowlers than it has been hitherto. Accommodation is cheap, living plentiful and wholesome, if plain, and any English gentleman desirous of a quiet shooting tour would find in the Colony numbers of farmers only too delighted to put him up and show him sport.

Within the Cape Colony alone there are to be found no less than five or six different kinds of francolin, six or seven kinds of bustard, two species of quail, one of the guinea-fowl, one or two of the sand-grouse family, and two sorts of snipe.

In this list I have not included wild duck, wild geese, teal, widgeon, and rails, many of the species of which abound. This is a goodly category, and even the most exacting of sportsmen would hardly be disposed to cavil at its proportions. Add to the ordinary attractions of game shooting a magnificent country, offering almost every conceivable variety of scenery, a climate always healthful, often perfect, and seldom too intemperate or too extreme for sport, and my humble contention on behalf of the claims of the Cape game birds is, I venture to believe, not difficult of substantiation.

I have witnessed some memorable days amongst the antelopes of the Colony, whether springbok, rhebok, klipspringer, duyker, bushbok, or steinbok—days never to be forgotten for their manifold and indescribable charms; but I think, amongst almost equally cherished recollections that I have, are some of the quiet days of sport enjoyed in some wild mountain kloof, some shaggy upland, or some broad and spreading karroo, wandering quietly in search of the red or grey-wing "partridge," the "pheasant" (all francolins, though called otherwise by the Cape settlers), the guinea-fowl, the Namaqua "partridge," or the magnificent bustards, all of which, in their different haunts, grace in plenty the landscapes of the Old Colony.

First among the Cape game birds I will place the francolins. Amongst these, the so-called "pheasant" of the colonists (*Francolinus clamator*) may first be taken. Commonly to be met with in the broad belt of bushy country that fringes the coast-line of the Colony, and extends to the eastward considerably into the interior, the "pheasant" may be recognised

SOME GAME BIRDS OF CAPE COLONY. I.

1. Namaqua Partridge (Sandgrouse), (Pterocles Tachypetes).
2. Guinea Fowl (Numida Mitrata).
3. Red-wing Partridge (Francolinus Le Vaillantii).
4. Cape Pheasant (Francolinus Clamator).
5. Black Koorhaan; a Bustard (Otis Atra).
6. Vaal (Grey) Koorhaan; a Bustar (Otis Vigorsii).

at great distances by its noisy, clamouring call. It is the largest of its genus, and is a fine game bird, measuring some fifteen or sixteen inches in length. Its general colour is a dark brown, but each feather is marked with narrow whitish lines, which concentrate at the shaft. The chin and throat are white, and the top of the head quite a blackish brown; the brown neck feathers have white edges, which impart a singular appearance to this bird; while the stomach and side feathers are slightly speckled with white, and have a broadish white line running down the centre. The only objection, that I am aware of, that can be urged against Cape game birds when in the field, is that they are often much more troublesome to flush than feathered game in England. They will sometimes lie absolutely like stones, and have to be kicked up; or, especially in the case of the pheasant and the guinea-fowl, they will run like hares. The first of these peculiarities will, to the moderate shot, not be thought altogether inconvenient, for a bird that rises at your feet, as a rule, is not difficult to account for, if your powder be anything like straight. But these francolins are, in one respect, even worse sinners than their brethren, for they have a very common habit of perching amongst bushes and low trees, and in dense jungle; they are, therefore, exceedingly difficult to drive out. Yet even with these drawbacks, if you do not object to early rising—and at the Cape this is a habit you soon learn to acquire—and will go forth at four o'clock, or thereabouts, you will certainly find the pheasants feeding upon open ground, and especially in the vicinity of water-courses or moist low-lying soil. Kolben, who sojourned at the Cape between

1704 and 1714, gravely informs us that these birds "will suffer a man, behind the picture of a pheasant, to approach near enough to throw a net over them." How Kolben picked up this extraordinary myth I am not aware. He appears, however, to have been of a confiding nature, and although he acquired much useful information concerning the country, he was made the recipient of many absurd notions, which he gravely related upon reaching Europe. Certain it is that the pheasants now lack the artistic sense claimed for them by Kolben, and are not to be taken by such a pictorial device in these days.

I first met with the pheasants upon a journey from Witte Poort, north of the Uitenhage district, to the coast. We had outspanned for the night in the bush-veldt country, about a day's journey from the town of Uitenhage, and as time was of no great importance, we rose at half-past three next morning, and proceeded to a broad, well-bushed kloof, having here and there considerable spaces of open ground and a small water-course running through the centre. Our first quarry chanced to be a bush buck (*Tragelaphus sylvaticus*), one of the handsomest and pluckiest of the Cape antelopes, which we surprised near the water, and secured with two charges of buckshot, with which we had been prepared, as we knew these buck abounded hereabouts. Half-a-mile farther on we came to a broad stretch of grassy open ground, whereon here and there a few low bushes grew, while near the water-course which intersected it the palmiet flourished luxuriantly, affording excellent cover. Upon this ground, and for some distance up the kloof, as we had expected, the pheasants and grey and red-wing partridges abounded, and for about two hours and a half we had capital shooting. It is true that the pheasants occasionally ran clean out of view, but on the other hand quite as many lay like stones, and after a few miles walking, our three guns had amassed a bag of nine brace of pheasants and partridges, besides a couple of the colonial wild duck (*Anas flavirostris*), called by the Boers the geelbec (yellow bill). I may add that the shooting was much more easy than if English partridges had been our game. We returned to the farmhouse, needless to state, in a contented frame of mind, and with enormous appetites. The francolins and the bustards of the Cape are good table birds, but they have, in common with most of the game of Africa, a certain dryness which renders them not quite such delicious eating as the game birds of our own temperate climate. Nevertheless, they are, as a rule, undeniably good eating, and the great kori bustard, which puts on fat in a quite amazing manner, is, in particular, fitted to grace the most princely banquet, or to satisfy the most exigent appetite.

 The red-necked francolin (*Francolinus nudicollis*), also called "pheasant," is of much the same size as *Francolinus clamator*. Its body is brown, the back feathers having black centres; the chest feathers are a lighter brown, having the same black marking; and the neck feathers are white, having broad black stripes. The breast, stomach, and flank feathers are of a darker brown, and have white stripes. The front of the throat is bare and of a bright red colour (from whence the bird, I suppose, takes its name), and the space round the eyes, and the chin, are also bare and bright red in colour. I shot a few of this fine francolin in some wooded kloofs near the eastern end of the Zwartberg, but they were not very abundant, and were difficult to get at. In the Knysna Forest, and in the forest country of the eastern frontier and in other wooded districts, they are abundant, and may be secured at early morning, while feeding, much in the same way as the common "pheasant." The *Francolinus adspersus* of Waterhouse, first discovered by Sir J. Alexander in his travels in 1835-6, a brownish grey bird, a foot in length, is, I believe, found on the southern banks of the Orange River in the north-west of the Colony, but very little is known of it.

The grey-wing francolin (*Francolinus afer*), here called "partridge," comes next. In general colour this bird is of a light ashy grey, marked upon the back with black blotches and reddish brown bars. The chest and flanks are blurred with darker reddish brown markings, and the stomach is thinly barred with dark brown. There are rufous bands down the head and neck, and under the bill to the chest. The chin and throat are white, and a white band runs from behind the eye to the shoulder; the tail is dark brown, marked with reddish brown bars. Its length averages about a foot, and it nests among bushes. This francolin is a really good game bird. It is flushed more readily than its congener and frequent companion, the red-wing partridge, which is sometimes found almost immovably fixed in its thick shelter, and in the early morning and evening it may generally be found upon open ground. I have enjoyed some excellent days' shooting with this francolin, especially in some of the shallower mountain kloofs, and upon the sides and tops of hills, surrounded on every side by scenery of surpassing beauty.

The red-wing francolin, or "partridge" (*Francolinus Le Vaillantii*), is, to my mind, taken all round, one of the handsomest and most representative of South African game birds. In size it slightly exceeds the grey-wing, but in colouring it far outvies that bird. In general colour it is not dissimilar, but the markings are much darker and brighter. Thin white stripes, mottled with black, run from the beak, over each eye, to the back of the head, where they unite and extend further. A similar stripe runs under the eye across the ear, trending to the chest, where it broadens into a crescent not unlike the white gorget of the ring ousel of this country, but bigger. A beautiful rich orange-red band surrounds the eye, passes over the ear, then widens and spreads backward to the neck and forward to the white crescent upon the chest. The front part of the throat is of the same rich colour. The chest beneath the white crescent, the stomach, and sides are beautifully coloured with dark rufous and brown mottlings, and the insides of the wing feathers are dark rufous—from whence the bird takes its name. In a deep, far extending kloof or glen, some miles from Naroekas Poort, through which meandered a small stream, these fine birds abounded, and here we often had very excellent sport with them. They lie, as I have before mentioned, often with extraordinary tenacity, and especially in the thick palmiet bordering the streams, and in long grass, they can sometimes hardly be forced to rise. In some localities they are found on higher ground, like the grey-wing partridge, but, as a rule, they are not often seen very far away from moist ground. By getting up betimes, and reaching their favourite ground in early morning, we could generally reckon upon a bag of from five to a dozen brace of these handsome francolins. From its habits, this bird is more easily shot than the grey-wing without the aid of a dog—if, as is unfortunately sometimes the case, such a convenience is not readily procurable.

The first mention of the two Cape partridges last before-mentioned seems to have been made by William Ten Rhyne, a native of Daventry, a physician, who afterwards became a member of the Council of Justice of the Dutch East India Company. Ten Rhyne made the voyage to the Cape as far back as 1673, and his remarks may be taken as amongst the earliest upon the natural history of Cape Colony. He mentions "Rubicundi et cinerei phasiani," which, being translated, and rendering "phàsiani" as partridges, I take, from the similarity of colouring, to be the earliest description of the red-wing and grey-wing francolins of the Cape.

The *Francolinus subtorquatus* of Sir A. Smith, a species rather smaller than the red-wing, is stated not to have occurred within the Cape Colony, although found in the Orange Free State and Natal; but, from what I have heard, I am by no means certain that it has not been found pretty frequently in the north of the Colony, not far from the southern banks of the Orange River.

The wild guinea-fowl (*Numida mitrata*) I will next treat of, in continuing this subject of the game birds. Although a much handsomer and finer bird than the common domesticated guinea-fowl of our own poultry yards, it is not very dissimilar in general appearance, save that its colour is darker, and the white mottlings are more noticeable. It is found gregariously and in abundance in the dense bush-veldt country near the Sunday and Great Fish Rivers, and upon the northern borders near the Orange River, as well as more sparingly in other bushy or semi-bushy localities. We shot some of these birds on our way up and down country between Uitenhage and Graaff Reinet; near the Sunday River they were especially plentiful, and, like all their kind, we found them feeble in flight, but great runners and roosters. In the evening they may be sometimes shot with a bullet while roosting in low trees near the water.

The sand-grouse family is represented in the Cape Colony by the well-known "Namaqua partridge," as it is called all over South Africa. This bird, the *Pterocles tachypetes* of Temminck, the Namaqua grouse of Shaw, and Namaqua patrys of the Boers, abounds upon every karroo plain of the Colony; and towards evening, when flying to water, may be often witnessed in great flocks, and numbers may then be shot. In general colour this bird is a dull ashy brown; the back is of a darker shade and mottled, while the wing feathers are dark brown. Between the chest and stomach are two bands, one white, the other reddish brown. The throat and chin are yellow. In length this bird extends to about eleven inches. Its long and pointed wings denote great powers of flight. The legs are very short, and the feet feathered, and when seen, as it often may be, running about the karroo, the aspect of this bird is very ludicrous; its gait has, indeed, been compared to the appearance of a clock-work mouse when set in motion. This sand-grouse seems likely to be abundant in the Colony for all time. It is not much shot for food, as its flesh is the driest of all South African game birds. It is, however, if included in a game stew, by no means bad eating, and as it affords very good shooting, new comers to the Colony will not, I am convinced, feel inclined to despise it. The sand-grouse will always find its way to water at close of day, and this very peculiarity has often saved the lives of men and beasts in many a parching trek, when the evening flight of this bird has indicated the presence of some unknown fountain in the desert.

Of bustards there are six or seven kinds to be found within the Colony. First and noblest of the family, and, indeed, of all South African feathered game, stands the great kori bustard, the gom paauw of the Dutch, the kori of the Bechuanas, the *Otis kori* of the naturalist, Dr. Burchell. Formerly abundant upon every open karroo, this magnificent bird is at the present day by no means so plentiful in the Colony; its great size, and quality as a table bird, having rendered it much sought after by sportsmen, whether Dutch, British, or native. It is, however, still to be met with occasionally upon the flats, generally not far from the mimosa bushes fringing rivers and dry water-courses, and as it is a migratory bird, it will probably long remain to the Colony, although in reduced numbers. The Dutch Afrikanders have ever loved to bestow quaint and outlandish names upon the birds and beasts of their adopted land. The name gom paauw really signifies gum peacock; this bustard being quaintly called by the Boers "peacock" from its fancied resemblance to that bird in its habits during the pairing and breeding season. This bustard and no other is supposed to have an inordinate affection for the gum which exudes from the mimosa thickets (near which it is often found) at certain seasons, and it then attains its extraordinary fatness and condition.

I first met with and shot the paauw between Jansenville and Graaff Reinet, and I shall never forget the exultation with which my shooting companion and myself witnessed the fall of the great bird to our rifle bullets at about sixty paces. This particular individual weighed twenty-six pounds, but they frequently run from thirty pounds to thirty-five pounds, and even forty pounds, and farther up in the interior they reach, I believe, the enormous weight of from sixty pounds to seventy pounds. During a lengthened stay in the Midland and Eastern provinces of Cape Colony, I saw only some half-dozen specimens of this noble bird, and of these I shot only one other, for they are exceedingly shy and difficult of approach. In some seasons they are, however, more plentiful, and especially when excessive droughts prevail beyond the Orange River. These bustards are often extraordinarily fat, so much so as to make one wonder how they can attain such adipose proportions in so hot and dry a climate. An old colonial friend, Mr. J. B. Evans, of Riet Fontein, on whose extensive farms I have often shot at the Cape, gave me a striking instance of this characteristic.

Years ago, when this bird was much more

plentiful on the karroo than it now is, he and a friend, while out springbok shooting, espied a huge paauw stalking over the plain in the direction of a rocky hill. The hunters turned their attention at once to the paauw, and after some manœuvring, drove it to the base of this hill. When within about seventy yards, just as they were thinking of firing, the bird rose and flew upwards towards the krantz or crown of the mountain. Both discharged their rifles, and by good fortune one of the bullets told. The effect was extraordinary. The mighty bird, sore stricken, instantly towered high in air above the hill, and then fell like a feathered thunderbolt. On reaching the spot where their game had fallen, the hunters, to their chagrin, found upon the rocks literally nothing but the scattered and pulpy remains of the paauw, which had burst to atoms. The bird must have been more than ordinarily fat, for the amount of adipose tissue lying about was beyond belief. Any Cape colonist can, however, bear testimony to the wonderfully high condition these birds attain. The general colour of the gom paauw is an ashen grey strongly mottled. The wings are lighter, and in some specimens almost white, and are black mottled. The wing feathers are black, the neck and head grey barred with black. The top of the head is black and heavily crested, and the stomach and chest are white, with a band of black separating the neck from the chest. The legs are yellow. An average cock bird will measure about four feet eight inches in length. The female is considerably smaller than the male, and her plumage less bright. The body of this bird is so robust, and so strongly protected by feathers, that shot make little impression upon it,

SOME GAME BIRDS OF CAPE COLONY. II.
1. Veldt Paauw (Eupodotis Caffra).   2. Gom Paauw, or Great Kori Bustard (Otis Kori).

especially as one seldom gets to very close quarters, and its death is invariably compassed by rifle and bullet. The Boers of old soon learned by experience to shoot them with their long roers (smooth-bore rifles), carrying large spherical bullets. The flesh is not altogether unlike that of the turkey; but it partakes also of the true game flavour, and is delicious eating.

The next in size of the bustards is the *Eupodotis caffra* of Lichtenstein, commonly called the veldt paauw by the colonists. This is a fine big bird, often measuring in length from three feet two inches to three feet four inches. Its general colour is dull ashen grey, streaked with black, and relieved by touches of white upon the chin, neck, and stomach, upon which there are white bands. It is more commonly found than its congener the gom paauw, and is pretty frequently seen and shot upon the Great Karroo, especially to the north and east. Upon the plains of Bushmanland it is also plentiful.

Another fine bustard is *Otis cœlei*, a bird measuring three feet six inches in length. Its colour is a yellowish brown, interspersed with fine dark brown lines. The chest, underparts, and front of the neck are white, tinged with bluish grey. It has a reddish coloured ruff, a white tail heavily barred with black, and the legs and bill are yellow. The females of both these last-named varieties are smaller than the males, and both varieties are excellent sporting birds, and equally good at table.

One of the best known bustards in the Colony is the knorhaan, commonly called the black koorhaan.*

---

* The old Dutch name knorhaan, literally "scolding cock," appears to have been corrupted to koorhaan, sometimes koran.

the *Otis atra* of Linnæus, or *Eupodotis afra* of Gmelin. The name knorhaan, given to it by the Dutch, literally signifying scolding cock, aptly describes it, for this is one of the noisiest, most provoking of birds. Its harsh discordant chattering "craak" completely disturbs the veldt in its immediate vicinity, and it is further a great runner. In spite of all this, the black koorhaan is a fine sporting bird, and when you know its little eccentricities and can master your impatience, affords really good sport, for it is exceedingly common upon every karroo of the Colony. It is of a brownish black colour, streaked with rufous. It has a white stripe from the eye to the base of the head, and a collar marking on the back of the neck, also of white, and there are distinctive white markings upon the wings. Here is its summary by Peter Kolben (*circa* 1704) : "A bird peculiar to the Cape countries is called the knorcock, the female knorhen. These serve like sentinels to the other fowls by a loud noise they make on seeing a man, which resembles the word 'crack,' and which they repeat very clamorously . . . is about the size of a hen." This bustard, as Kolben further notices, is not a great flyer, and if flushed very soon goes down again, when it watches its pursuer with the greatest pertinacity. The best plan to work its downfall is to approach it from different directions, when its attention is distracted from one gunner to the other. The vaal or grey koorhaan (*Otis vigorsii*) is of a greyish red colour, mottled with black and dark brown. The head is very slightly crested, and the bird has a peculiarly beautiful pinkish gloss upon its plumage, which fades after death. In size it much resembles the common koorhaan, measuring some nineteen or twenty inches in length, and it is abundant on the karroo, though not quite so commonly met with as the black koorhaan. This bird has a habit of squatting very closely, and can be frequently approached pretty easily in consequence. It is a good sporting and table bird.

Perhaps the most beautiful of the smaller Cape bustards is the blue koorhaan, *Eupodotis cœrulescens*, a bird not found so generally distributed as the two last-named. I came across it occasionally in the Eastern province in open or sparsely bushed country. Its colour is mainly blue, the top of the head, back and front of neck, and the breast and stomach being all of that colour. The upper part of the throat and the forehead are black, and the rest of the bird is of a reddish brown colour. The legs are yellow, and there are one or two white markings round and upon the top of the head. This is a singularly striking bird, but it is somewhat capriciously distributed, mainly in the Eastern province. When once found, however, it is fairly plentiful, and gives capital sport.

One other bustard (the *Eupodotis afroides* of Sir Andrew Smith) concludes the list. This species, differing very little from the common black koorhaan, is occasionally met with just south of the Orange River, but can hardly be classed as one of the ordinary sporting birds of the Colony.

The common quail (*Coturnix dactylisonans* of Temminck) is so well known as a game bird as to require no description. It arrives at the Cape towards the end of August, and may be found abundantly in the Western province. Bags of from twenty to thirty brace to one gun are not uncommon.

I have enjoyed days of delightful shooting with this little beauty when staying at Cape Town, after driving some distance out into the country to the shooting ground of the friends who kindly invited me.

The sand-quail or reed-quail of the colonists (*Turnix hottentotus*), belonging to the sub-family of bush-quails, is much less abundant than the common quail, and is only to be met with occasionally in the maritime districts of the Colony, principally to the south and south-west. This bird never quits the Colony, and breeds near vleys and reedy places. It never flies far, is a great runner, and is generally to be found in thick covert; in size, it is a trifle smaller than the last-named.

The two snipes to be found at the Cape must conclude my list of sporting birds. Of these, the commoner is *Gallinago æquatorialis*, which may be called the Cape snipe, and is to be found in every part of the Colony where water or swampy ground exists. I believe this snipe has been and is still mistaken for the common snipe, which, however, it exceeds slightly in size; it is also much blacker on the back.

The painted snipe (*Rhynchæa capensis*) is often found in abundance in the same moist localities as the last-named. Its colouring is very remarkable. The brown head is lightly touched with white. Over the centre of the head a yellow stripe runs from the base of the bill to the back of the neck, and another similar stripe runs from either eye to the back of the head, while yet two other yellow stripes, edged with black, trend from the shoulders, down the back, to the base of the tail. The brown neck has a white collar; the under parts of the bird are white, and the back is barred with black, grey, and white. The wings are variegated with black and yellow spots or eyes, and, with the tail, have black wavy cross bars. This beautiful bird is slightly smaller than the common Cape snipe, and the female is, curiously enough, even more beautifully coloured than the male. The eye stripe in the hen bird is white instead of yellow, and the neck and breast are reddish brown instead of grey brown ; the wings are greenish.

In addition to this considerable catalogue of sporting birds, there are rails, plovers, wild geese, widgeon, wild duck, and teal to be found abundantly in various parts of the Cape Colony.

Wanderers in search of health or sport, or of health and sport combined, will find, within nineteen days of our own shores, a country rich in scenes of changing and romantic beauty, and one of the most health-giving climates of the world; rich—wondrously rich—in its flora and fauna. One only drawback I know of—the average accommodation is, though sufficing for rough-and-ready sportsmen, at present primitive, and at times rude. But even in this respect the Colony is improving, and it will improve more rapidly as demands increase.

CHAPTER XVIII.

THE BOER OF TO-DAY.

IN many respects the Boer of the present day exhibits but little change from his sturdy forefathers of 200 years ago. He sees so little of the busy outer world, of which he has no conception, and no desire to know; and surrounded by his usually large family and his flocks and herds, he plods dully along his life road, whether he be wine grower, or pastoral farmer, or Trek, or wandering Boer, with no wish to better or change his lot. In the Cape Colony one day's routine exactly resembles another. The family rise at dawn, and after a cup of coffee and devotions, the flocks are unkraaled, and, in charge of Hottentot or Kaffir boys, are despatched to their pasture on the veldt or mountain until the evening. If the Boer is a wine grower, the work of the vineyard is directed, and tobacco and fruit crops are looked after, and other matters put in train for the day. At eight o'clock breakfast is eaten, and at mid-day, dinner. Then follows a siesta, and at sunset the horses, cattle, sheep, and goats are put into kraal for the night. Then come the evening meal and family devotions, and at eight o'clock, or at latest nine, all are asleep. Smoking is vigorously indulged in during the whole of the day, but this is not an expensive luxury, as Boer tobacco costs little or nothing to produce, and is sold at a retail price of from twopence to sixpence per pound. Large quantities of coffee are also consumed, and, strangely enough, sweets are a favourite luxury among the Boers. The Bible is the only book that is generally read, and twice a year at least the family trek into the nearest town to partake of the Nacht-maal or Sacrament. The Boer never changes his clothes for his night's rest, and his ablutions are scanty enough; but these habits have descended to him from his forefathers, who, from the manifold dangers and night alarms of the wilderness, were formerly compelled to sleep in their clothes, so as to be able to go out at a moment's notice when danger threatened their flocks or themselves.

In the Western province, the oldest of the Dutch settlements, some few refinements are creeping into the dwelling-houses. Some of the daughters may have been educated at the ladies' schools now established in various parts of the Colony, and the notes of a harmonium, or even a piano, may now be occasionally heard in a few of the better houses. But with the Boers progress is so slow, and their hatred of innovations so rooted, that many years may elapse before even the simplest refinements of civilisation acquire much foothold.

The vrouw occupies an important position in every Dutch household, as, indeed, she ought. In the stirring history of the Boers in South Africa, the womenfolk have ever borne their share, and more than their share of the work, rousing their phlegmatic spouses at need by sharp and scathing words, and in time of battle often assisting to repel the attacks of savage foes, loading the emptied rifles of their men-kind, and, upon occasion, even using them themselves. If the secret history of the Transvaal

War could be written, the part played by the housewives in stirring up the spirit of discontent and insurrection, would not improbably be the cause of some wonderment and even admiration; for in restless determination the Dutch Afrikander women are often much in advance of their stolid husbands. The *huis-vrouw* may be found usually from dawn till nightfall, seated in her easy-chair, directing the polity of the establishment. If the weather is chilly, as in spring and winter it often is, her ample feet not seldom repose upon a kind of footstool filled with hot charcoal. No other person would dare to occupy the vrouw's seat of honour under any circumstances whatever. In the remote districts, and farther up in the Transvaal and Free State, many of the Boers are to this day as primitive in their habits as when Barrow wrote of the Dutch housewife more than ninety years ago. "She makes no scruple of having her legs and feet washed in warm water by a slave before strangers. . . If the motive of such a custom were that of cleanliness, the practice of it would deserve praise; but to see the tub with the same water passed round through all the branches of the family, according to seniority, is apt to create ideas of a very different nature." This evening custom still obtains in the households of the remote back country, and to the stranger, who is expected to share the luxury, is rather appalling. From her huge arm-chair the vrouw exercises a severe sway over her numerous native servants, only rising occasionally to bake bread, make coffee, and superintend the butchering department, and the manufacture of dried flesh, soaps, candles, and other necessaries. Occasionally, if the servants are away or otherwise employed, the lady will slaughter and cut up a sheep or goat with her own hands, and a capital business she makes of it; indeed, with some of these simple pastoral people, butchering appears to have a great attraction among the fair sex. It is not a pleasant *trait*, but a really useful one.

The morning ablutions are very scanty. A Hottentot appears with a basin of minute dimensions, containing, perhaps, a pint or so of water, and a small and often dirty towel. The father dips his fingers in the water, gives his eyes and a small portion of his face a slight rub, dries himself on the towel, and the trick is done. Then the process is repeated by the rest of the family, all using the same water and the same towel. I have watched this proceeding with extreme curiosity, tempered with some little disgust, in several Boer establishments, and I am given to understand that no other ablutions whatever are thought requisite, year in, year out—save, of course, the evening custom I have above referred to. But then water is often a very scarce luxury in South Africa, and the Dutch Afrikander is at least 200 years behind the times. Our own grandfathers troubled the washstand very little, if we may judge from the size of basins of a hundred years ago. Prayers are never forgotten, and are conducted with the most solemn ceremony. The father reads a chapter from the great family clasp-bible, which has often been in the family a hundred years or more, and even sometimes ever since Van Riebeek's time (1652). Then, all kneeling, prayers are offered up; the Boer prays for himself, his family, and his flocks, and other earthly possessions—if in the Transvaal or Orange Free State, for the President and *his* family and worldly gear—as well as for the spiritual welfare in either case. At meals a long grace is usually said before falling to.

Few people are more litigious than the Cape Dutch, considering the immense distances that often separate them from their legal advisers. But, disputatious though they are, it is unquestionable that on occasion they altogether surpass their British fellow-subjects in the strength and completeness of their apologies. Here is a public apology, translated from the columns of a Cape paper, *Di Afrikaanse Patriot*:—

"I, the undersigned, A. C. du Plessis, C. son, retract hereby everything I have said against the innocent Mr. G. P. Bezuidenhout, calling myself an infamous liar, and striking my mouth with the exclamation, 'You mendacious mouth (jij lengenachtige bek), why do you lie so?' I declare further that I know nothing against the character of Mr. G. P. Bezuidenhout. I call myself, besides, a genuine liar of the first class. (Signed) A. C. du Plessis. Witnesses, P. du Plessis, J. C. Holmes."

I have said that the Cape Dutch are great smokers; they are equally great spitters, and herein they vie with the most notable talent of America; this habit they even carry with them to the church. A new Dutch Reformed Church was recently built, and the *Graaff Reinetter*, a local paper, was moved to speak thus in view of preserving the virgin purity of the floor—made of pitch pine and oiled. Here is a translation of the article in question :—

"In the last meeting of the congregation the question was discussed whether it was necessary to have spittoons made for the new church in places where persons befoul the floor by spitting on it. A few members of the congregation thought that those who wanted spittoons could provide themselves with them. The chairman said that the complaint had been recently made that a man and his wife got quite sick from this spitting near them; even the Sunday school teachers, who taught their classes on the gallery of the old church, had complained that there were so many quids (*pruimpjes*) and so much tobacco spittle on the floor, that they got quite a turn in their stomachs. Some of the speakers thought that the supplying of spittoons would only make the spitting worse. Generally, the meeting came to the conclusion that they hoped no one would be dirty enough to befoul the floor of our fine new church with spittle or quids, to the disedification of decent hearers and the defilement of God's house."

The article went on to compare the spitters to Kaffirs, and to ask "how one can give heed to the Word, and heartily sing or join in prayer, while his attention is divided between his plug of tobacco and the service."

The average size of an ordinary Cape farm is about 3,000 morgen—rather more than 6,000 acres. For a large pastoral farmer, this is a mere fleabite, and farms of 20,000 and 40,000 acres, and even more, are very commonly met with. But it is not to be supposed that all Boers in the Colony are great landed proprietors. There yet remain many who move about from place to place with their flocks and herds, sometimes paying a trifle (in kind) for pasturage and water, oftener finding it for nothing upon the vast crown lands, that still lie unused, in millions of acres, in many parts of the Colony. Many of these families have never known a home, save their own waggons; they have been born and reared in this fashion, and so they bring forth and rear their own children. A strange existence, indeed! These nomads pay no taxes, and, practically, own no laws, and by all careful and prudent farmers, they are looked upon with unmixed disgust and resentment. It is probable that a great part of the contagious diseases that ravage the flocks and herds of the colonial farmers—such as scab, brand-sickte, red-water, and lung-sickness — are transmitted by these Trek Boers. Theirs is, no doubt, a picturesque and interesting survival of the old wandering existence of the early Dutch settlers, but it is a survival not exactly fitted for these days of progress, and its departure would greatly benefit the Colony.

In the north-west divisions of the Colony, notably in Calvinia, Namaqualand, Fraserburg, and Victoria West, even the well-to-do farmers move away, periodically, from their regular homes, and with families and flocks, seek fresh pasturage in the trek-veldt (part of the ancient Bushmanland), where their white waggon tilts and camping tents lie dotted upon the landscape in picturesque disorder.

The curious customs of the young Boers when courting have too often been touched upon to need repetition here. The smart attire, the shining riding boots, the bright new spurs, the curvetting horse, the "opsitting," and the "opsitting" candle, by the burning of which the duration of the lovers' interview is regulated, are all pretty well known. The following Boer Dutch verses, composed by an old Cape friend of mine,* very well express the courting preparations:

> "Carl Jansen was een jonge kerel,
>   En dapper ook was hij.
> Fluks was hij met zyn laange roer,
>   En ook met paarde reij.
>
> Carl Jansen klim zyn spoght piart op,
>   De beste wat hij kon krij.
> Dan gaat hij voort al lynks en reghts,
>   Met de meisjes o'r als te vrej.
>
> Carl Jansen komt na' onze huis,
>   Hij jaa moss, oer de vlak.
> Zyn stevels het al mooi geschine,
>   En een kerse was en zyn zaak."

Even in the long-settled Cape Colony, Boer husbandry is too frequently as primitive as it was two centuries since. The Boer has ever preferred pastoral farming, with its loose roving methods, and free-and-easy camp life in the trek-veldt for part of the year, to agriculture. Far greater returns might be obtained from the soil, if a little more trouble were taken; if the ground were a little better prepared, and more care were taken to plough deeper and more evenly. As often as not in the back settlements, the bushes and roots are not cleared away, nor the clods broken by rolling and harrowing. It is an astonishing fact, that thrashing is performed in most parts of the Colony, in the ancient Biblical manner, by treading out the corn on a circular threshing floor by means of horses or oxen; the corn is then cleared of the chaff, simply by throwing it in the air when a good breeze is blowing, and it is then bagged and stored away ready for sale or use. Thrashing, which by the aid of machinery

---

* Mr. A. G. Evans, an eighteen years' resident among the Boers of Sneeuwberg.

could be finished in a week or ten days, often lasts from January to June, after a quite astonishing amount of labour.

The Boer is just beginning to find out that banks are really sound and reliable institutions, not intended for the robbery of the farmer of his hardly won money. But even now very few take advantage of the bank in the nearest country town; they cannot be brought to believe that any man can be so foolish as to pay them interest for the privilege of taking charge of their savings, and rather look upon this as a cunning trap, devised to snare the unwary. The average up-country Boer prefers to accumulate his savings in a chest in the bedroom, and amongst the richer members of the community, it was not an uncommon thing, in the good days, to have £10,000 or £12,000 lying in the house in specie. Even now, despite recent bad times, large hoards lie stored in the family chest.

Stories of the Cape Dutch and their quaint ways are innumerable; here is one of them. Before the Cape railway to Wellington was completed, a Boer transport rider used to drive his ox-waggon, heavily laden with produce, over the deep and distressing sands that form the Cape Flats, and at this point often had to call in the aid of a second span of oxen.

When the railway was built, the old fellow was tempted, with some heart-quaking, to try a journey by train over his old trekking-ground. As sometimes happened then on Cape railways, the engine had to pull up on these very flats to get up more steam. Thereupon the old Dutchman put his head out of the carriage window, and shouted, " Ik doch zo blas als jij bei die dek zand kom " (" I thought you would want a blow when you got to the deep sand "). The engine was puffing away at a great rate, and the old fellow, who could scarcely be persuaded to look upon it as an inanimate thing, seemed to have thought that its stoppage was due to the same want of breath that formerly gave pause to his own oxen on this selfsame spot.

Of all races of mankind, the Boers have in time past been endowed with the richest and most diversified hunting grounds that the world has produced. On the arrival of the father settlers at the Cape, they found a land which for untold centuries had teemed with wild animals of every conceivable size and shape, from the mammoth elephant to the tiniest of those twenty-five odd varieties of antelope which South Africa boasts of. The elephant, the rhinoceros, and the hippopotamus were then plentiful over what is now the Cape Colony, and all the larger antelopes, such as the eland, roan antelope, and others, have only been driven from the Colony itself within the memory of man. It must be admitted that the Boers have not been slow to avail themselves of Nature's profusion. Since 1652, when lions wandered through the gardens of stout old Governor Van Riebeek, on the ground where Cape Town now stands, the Dutch have waged an incessant and a frightfully wasteful warfare upon game of all descriptions. Within the last fifteen or twenty years, since they have availed themselves of breech-loading weapons, the decimation of antelopes, elephants, rhinoceroses, zebras, and quaggas, has been prodigious. Antelopes were slaughtered in thousands upon thousands for the sake of their skins, which were sold at very low prices, and shipped to England and elsewhere; and the plains of the Free State and Transvaal, which not long since literally swarmed with noble game, are now more denuded than even those of the Cape Colony. The rhinoceros is now nearly extinct south of the Zambesi, and the elephant only lingers in the most inaccessible fastnesses of the tsetse fly country, in Mashunaland, and one or two other localities. The smaller antelopes, it is true, are still plentiful, but only for the reason that their hides were not worth powder and ball. If much of the romance and pleasure of his life has departed from the Boer with the decrease of game, he has only himself to thank.

Of late years the dress of the Boer has changed. The springbok, blessbok, and other antelopes which formerly grazed in countless thousands on the karroos of the Old Colony, the rolling flats of the Free State, and the high veldt of the Transvaal, from whose skins the garments of the farmer were made, have been so depleted and exterminated, that most Boers now dress in fustian or moleskin. "Veldt-schoons," however, shoes of home-tanned leather, with the hair outside, are still invariably used. True, the elephant hunters, who yet pursue their dangerous calling in the far interior of Mashunaland, the Mababē veldt, or the Zambesi regions, and the Vee, or pastoral Boers (literally stock farmers)—nomads, who wander as their fancy or fresh pastures may impel them—still wear the leathern garb of their ancestors; but ivory is not now so difficult to procure, and land so much more scarce and less easily "jumped," that these classes are not nearly so numerous as in former days. The Boers do not take kindly to trading, which is almost entirely in the hands of English, Scotch, or Afrikanders other than Dutch. They trek about the country, however, with their families, and flocks, and herds, when they wish to sell their horses and flocks, or have oranges, tobacco, brandy, and dried fruits to dispose of. They have entirely discarded the old-fashioned "roer" (heavy smooth-bore gun), and are now almost invariably armed with good breech-loading sporting rifles, and from their constant practice at game shooting are excellent shots, as Majuba Hill and other fields can testify. The men are usually tall, heavy and ungainly in their movements, and somewhat sullen of aspect, unless of Huguenot extraction, when they may be easily picked out. The Transvaal Boers very generally live part of the year on the high veldt or uplands, and in the winter or healthy season migrate with their entire families and belongings into the lower or bush-veldt country, where there is good pasturage, and where they live a happy gipsy kind of life for some months.

Ten years ago it appeared almost as if the ancient embittered feeling of the Dutch towards the British in South Africa, softened by years of peace, was likely to disappear.

Ten years ago we in the Old Colony, Boers and British, had made up our minds to forget our differences, shake hands, and pull together; and for years before that the Dutch had ever looked upon us with respect, tinged perhaps with a little wholesome awe—qualities long fixed in their dull natures by the ancient prestige of British arms. The lessons taught by Sir Harry Smith at Boomplaats, on the Free State plains, and by the suppression of the Van Jaarsfeld Insurrection in 1798, were still remembered. Suddenly, like a thunderclap, came the news of Majuba Hill, and that subsequent surrender everlastingly disgraceful to England's name. Since that day much is changed, and the Boer, whether he be colonial or a dweller in the republics beyond the Orange River, goes his way with altered demeanour. In the Old Colony, the Afrikander bond has done much to educate the Dutch party, and instead of viewing the elections with apathy, the up-country Boer now controls the helm of state, and, through his mouthpieces—Mr. Hofmeyr and others—dictates his own terms to the Premier for the time being of the Cape Parliament. Agitation skilfully conducted has proved as great a power among these rude people as among the newspaper-reading artisans of English cities. Only a short time back, in our district of Zwart Ruggens, a meeting of farmers— English and Dutch—was held, and among the subjects discussed was the advisability of supporting an Act for the compulsory fencing of farms, proposed by the Cape Legislature. An old Boer of our neighbourhood, Hans Knoetze by name, presently got up and passionately exclaimed, "Ik schiet moors dood die erste man wat durf een gut mark op myn grond voor een draat henning" (" I'll shoot as dead as mutton the first man who will dare to make a hole on my ground for fencing"). Before Majuba this old man would never have dared to make such a speech.

But, compared with their relations of the Transvaal, the Boers of the Old Colony are peaceable and harmless enough, and, indeed, on the whole, loyal subjects of the British crown, and, although Majuba rendered them for a time somewhat more aggressive than of yore, it is in the last degree unlikely that our rule at the Cape will ever be endangered by Dutch disturbance. The vast extension of influence acquired last year (1888) by the British in the Zambesi regions, and the ever-increasing flow of British emigration to the Transvaal, have happily put matters beyond all possibility of doubt, and the British colonist now sleeps with a far lighter heart than for the few years immediately succeeding the miserable Transvaal War.

Mahomet has said that "Paradise is under the shadow of swords." It may with truth be said of the Transvaal Boers, as of the colonial Boers of the Eastern frontier before the great trek of 1836, that, since that time, their paradise has been beneath the shadow of their trusty rifles. With the cumbrous elephant "roer" of a past generation, and with the breech-loading Winchester, or Westley-Richards, or Martini-Henry of the present day, they have well-nigh exterminated the great game of their country, have harried the native tribes within and without their borders, and, lastly, extracted from Mr. Gladstone, in 1881, a peace, in all respects glorious to themselves, and inglorious to the British. As I shall show hereafter, a little tact, a little good management, might have saved the Transvaal to the English without a blow, without the loss of a single life, Dutch or British. But the fates and Mr. Gladstone willed that a surrender, the blackest and most disgraceful that ever darkened the annals of this country, should be made at the muzzles of the rifles of these Transvaal farmers. Just one hundred years before, the capitulations of Burgoyne and Cornwallis, in the American War of Independence, had cast a shadow upon the fair page of English history; but, compared with the Majuba surrender, the failures of Burgoyne and Cornwallis were honourable incidents. These generals were then far separated from their home government, and were compelled in a great emergency to act upon their own responsibility and their own judgment. In 1881, every disgraceful item of surrender was flashed by wire direct to the scene of battle. The thing is done, but the British colonists, the Cape Dutch (in a different sense), and, too, the natives of South Africa, never will, never can forget the humiliating and exasperating dishonour that the English then suffered. Well might Earl Cairns say, in a memorable speech made in debate at that period:

"In all the ills we ever bore
We grieved, we wept, we never blushed before."

The Boers have been ever proud of their history in South Africa; and, indeed, little known though their achievements are, the story of their struggles for existence in that country, of their " trekking " from British rule, and of their bloody but victorious strife with Dingaan in Natal, and Moselikatse in the Transvaal, is so abounding in romantic episodes as to be scarcely credited in modern history. The fatal day of Majuba Hill, and the English reverses before that fight, have added a proud chapter to Dutch Afrikander history.

What is to be the future of the Dutch in South Africa is even now, despite the struggles and intrigues of the Afrikander Bond, still uncertain. That this strange people will remain rooted to South African soil is quite certain, but whether the Transvaal Dutch will suffer themselves to be swamped out of all national recognition by the British element is another matter. President Kruger incessantly warns his people to "keep up their shooting," and although it is improbable that Boer and British will ever again come to blows, it is yet uncertain how the Transvaal Dutch propose to preserve their fast disappearing "Republic of isolation." It is possible that large numbers, disgusted with the overcrowding of the gold-seekers, may again trek north, and seek once more new homes in some fresh and far distant land.

From their constant habit of warfare upon wild beasts and savage tribes, the Transvaalers, especially the younger of them, have acquired a more truculent and aggressive nature than the Cape colonists, and even than their more staid brethren of the Orange Free State. It is probable, however, that time and more peaceful years, and above all, that desire of gold just awakening among them, will in the future soften and temper these qualities, and that the Transvaal Dutch may, twenty years hence, be a much more changed race than the Cape colonists, over whom the tide of emigration has swept unheeding. It must never be forgotten, too, that large numbers of Transvaal Boers favoured the British occupation, and infinitely preferred British rule to their own halting and bankrupt government, existing prior to 1877.

Not very long since I met an old Cape friend, with whom I had sojourned formerly on his karroo farms in Cape Colony. After a long separation a long talk naturally ensued. One topic led to another, until we happened upon that inevitable one of ever-burning interest to all British at the Cape— the Boer War of 1880-81. "Well," I said, "and how are the Boers in the Old Colony behaving now; they are less 'cheeky' I hope by this time?" "Yes," said my friend, "they are, somewhat; and I think now that matters will settle down again, especially as Government is acting more firmly and a Zambesi Protectorate has been practically declared. By-the-bye, you remember Swanepoels Poort very well, don't you?—the place where you and your friends nearly came to grief at that nasty drift (ford) the dark night you first crossed it? Well, one afternoon, some little time since, I was coming out of the mountains that very way home, when I met a Boer outspanned at Stols's farmhouse. His waggon looked worn and dirty, and his oxen weary from long travel. He turned out to be a Transvaal Boer on his way into Oudtshoorn division to visit his numerous friends and relations. (It must be remembered that the Boers of the Transvaal and Cape Colony are almost all united by ties of family and of sympathy.) I rode up to his waggon and off-saddled to have a chat, and I found the man a very sensible well-informed Boer—indeed, of quite a superior class. 'Well,' I said to him, after talking of various matters, 'and how was it that you came to fight the British Government?' 'Allemagtig!' (Almighty) said the Dutchman; 'believe me, most of the Boers up there (pointing north-east)—the most respectable, I mean—who had a stake in the land to lose, were not a bit sorry when you English took over the country. But let me tell you one thing: your Government went quite the wrong way to work. Now if they had made maaters (chums) with Oom Paul (Kruger) and a few other of our leading men and given them posts, and if they had listened a little to them, and especially if Colonel Lanyon and others had not been so terribly hoogmoedaag (high and mighty), all would have gone well. And I will further say that I and numbers of farmers—indeed, the majority of them—would, if the proposals of our leaders had been given ear to, have been British subjects to-day—ay, and peaceable ones to boot. But your administrators and colonels and captains all rode too high a horse. Instead of conciliating they threatened; instead of trying to understand us, they laughed at us. You must understand, too, there were always among us numbers of freebooting and disaffected men—those with nothing to lose and everything to gain by disturbance; therefore, immediately these saw an opening they began to agitate; after which Kruger, Joubert, and other disappointed men were soon persuaded to join as leaders. And then, as you know, the gathering at Paard Kraal was got together, and shortly the war began. Naturally I, like many another, when war was proclaimed, dare not refuse to join my countrymen, although I wanted no fighting; if I had done so ruin would have fallen upon me, my flocks and farms would have been confiscated, and I should have left heeltemaal kaal (quite naked). Well, I joined the commando, and went to Laing's Nek, Ingogo, and Majuba. You have heard the story of Majuba Hill, of course? Allemagtig! I tell you it was astounding, and our baas-raaking (thrashing) your English rooibaatjes (red-jackets) was even more wonderful to us than to them. Your men were badly led, and their shooting was alte sclecht (most wretched). I never saw anything like the way they threw away the battle. It happened in this wise. About daybreak we in our camp down below were awakened to find the soldiers crowning the mountain. Then about eighty of us went up to meet them; and so creeping from bush to bush, from rock to rock, we slowly made our way upwards. We could see the rooibaatjes' heads sticking out over the tops of the high rocks, and, too, the flash and smoke of their rifles, which were discharged very rapidly and at random. But they fired anyhow, at least fifteen or twenty feet over our heads. Ja! die coegels het vere bookaut onz gevluit (Yes, the bullets whistled high above us). As we crept closer we picked off the poor soldiers, as they showed their heads, like so many dassies (rock-rabbits). Even then had they charged us with the bayonet they might have driven us back; but no doubt they thought us much more numerous than we actually were, and feared, too, our rifle-fire. Well, nearer and nearer we becrept them, until at last we saw them making some attempt to charge. Allemagtig! Poor General Colley, he was indeed a brave man—too brave, in fact; for almost at the instant he came forward, cheering his men, and calling on them to charge, he fell riddled by our bullets—stukken geschiet (shot to pieces). When that happened, then your men fled, and we shot them down like rheboks as they ran down the mountain-side; for our battle-blood was up. Many of them jumped down krantzes (rocks) of fearful height, by which some were severely injured, others killed moors-dood (stone-dead). I could not help but feel jammer (sorry) for the arme karels (poor fellows). To me it was a wonderful thing indeed; and to this day I cannot understand how we won the battle. Your rooi-baatjes are brave enough—at least, they were at Laing's Nek and Ingogo River; but what is the use in these days of breech-loaders, if those that hold them are as brave as lions and yet cannot aim straight and judge distances? As you know, we Boers are, so to say, born with rifles in our hands, and learn to shoot straight; and from infancy, climbing as we do the hills and crossing the plains in search of game, we have a continual training which enables us to judge distances to a nicety.'"

Such was the plain but interesting narrative given to me by a Cape colonist, an Englishman having, from thirty-five years' experience, a very intimate knowledge of the Boers of South Africa. According to the view of my friend's Dutch informant, if a little more forbearance, a little more tact, a little diplomacy, had been exercised towards the Transvaal Dutch and their leaders from 1877 to 1880; if a few administrative posts had been judiciously distributed, the Boer War would never have happened, and the Transvaal, with all its potentialities of wealth, would have remained to this day flourishing and contented beneath the British flag.

But on the other hand, I am bound to say a widely different rendering has been placed before me. I have been informed by a Cape gentleman, who held a high administrative post in the Transvaal during British rule, and whose opportunities of forming accurate judgment were unrivalled, that nothing would have ever induced Kruger, Joubert, and others of the Boer leaders to join hands and work with the English. Their hatred and hostility never could have been overcome by peaceful means. When Sir Bartle Frere went up to Pretoria, he repeatedly interviewed the Dutch irreconcilables, and offered them every possible concession short of positive rendition of the country. No man found his way so straitly to the hearts, even of the rugged Boers, as Sir Bartle Frere; but even he could not turn these men from their purpose. "Sir Bartol Ferreira," said Kruger, in his curious Dutch-English, at one of these interviews, "we want our country back; we will take nothing less than our country, and we will not rest night or day till we get it."

At length even the gentle and patient Frere gave up his efforts in despair, and Kruger and his supporters went their way still breathing threatenings and hostility against the English. Yet even with these men, a firm policy, supported by a strong home Government, would in the end have triumphed. Two reverses for English arms, and the crowning mercy (for the Boers) of Majuba sufficed, however, to turn the faint hearts of Lord Kimberley and Mr. Gladstone, and once more the magnificent Transvaal country reverted to Dutch authority.

It has been recently said that the Transvaal authorities were still coquetting with Germany, and that a Protectorate by that Empire over the South African Republic would be the outcome. This view I take leave entirely to differ from. The bulk of the Transvaal people might, and would with judicious management, have come peaceably and quietly under the British flag in 1877, or soon after, and having once come willingly they would have remained for ever British subjects. I was conversing not long since with a Dutch gentleman, formerly holding office in the Transvaal Volksraad before 1877, who assured me that if President Burgers had not been in such desperate haste to quit the sinking ship of the Transvaal State, and to secure the British pension he afterwards enjoyed, and that if Sir T. Shepstone had only waited six months longer, the ruin of the bankrupt Republic would have been complete, and all the inhabitants would have freely and gladly placed themselves under British rule. Britannia seized the plum before it was yet quite ripe; in another few months it would have fallen into her mouth. But since 1877 many things have happened. By the war of 1880-81, the pride and enthusiasm of the Transvaalers have been increased a thousand-fold.

The opening up of the gold-fields, and the influx of the mining element, have placed the exchequer of the new Republic far beyond the sordid cares and troubles that formerly beset it, and it is, in my judgment, in the last degree improbable that the sturdy Transvaal Boers will ever bow the knee to Germany; I believe, rather, that if put to extremity, they would range themselves with the British. The success of the Republic is now assured, but looking at the enormous increase of the British element there—in the next few years to be still more recruited—looking at the approach of railways through the Free State and Bechuanaland (now going steadily forward), and the wonderful changes, in every instance favourable to British interests, everywhere taking place south of the Zambesi, it is, I think, incontestable that the future of South Africa is assuredly now to be British and not Dutch. The two races are at length just beginning to move together in the march of progress. The dream of an United South Africa, owning allegiance to Great Britain, is not now an Utopian one, and ten years hence may see it peacefully accomplished. But, whatever the issue, I have no fear that the Boer will ever vanish from South Africa. I, for one, should be sorry indeed to see him go, or his old-world characteristics entirely swamped by the thrusting modern European element. As a social and historical study, this farmer of the African wilderness, shut off from his fellows by 200 years of a rude and semi-barbarous existence, is unique, and deeply interesting. That he has been able to retain the good qualities—and they are not few —that he possesses, a deep if simple faith, a wonderful power of self-help and self-reliance, an intense love of family, extraordinary skill as a hunter, no little knowledge as a grazier, and when put to it, an undaunted courage—speaks volumes for the better type of the South African Boer. Through evil and good report, through petty wars and perils innumerable, in sunshine as in storm, ill-equipped, and ever relying upon his own stout arm alone, he has clung to his beloved "Zud Afrika," his land of promise; and his magnificent faith in that wonderful country is, as we may now all see, entirely justified, and not one whit too greatly rewarded.

Chapter XIX.

THE RISE AND FALL OF UPINGTONIA.

A Frontier Drama.

SOME few years ago an expedition set out from the north-west border of the Transvaal, which attracted great local attention at the time. On the banks of the Limpopo, or Crocodile River, there lay, outspanned, many a stout waggon and many a yoke of oxen, which carried some seventy or eighty families of Dopper Boers, with their goods and fortunes. The Doppers are a sect holding religious views more severe and savage even than those of the Calvinists of old. They look upon the natives in exactly the same light as did the wandering Israelites, as mere hewers of wood and drawers of water, and they treat them invariably with merciless severity and even barbarity, for they truly believe that they have "the heathen for an inheritance." These hardy trekkers were proceeding in search of what they called "the promised land," which was in reality a rich territory in Ovampoland, far to the north-west of Lake N'Gami and the river systems of that region.

A year or so passed by after their departure, and at length fitful news began to leak through from the interior of sad disasters to those Boer trekkers. It became known, finally, that after long and weary journeying through the arid doorst-land (thirst-land), the expedition had been overtaken by the unhealthy season in the lonely and malarious swamp country lying between the rivers Teouge and Chobe; that their oxen had nearly all perished from tsetse fly and lung sickness; that whole families had been carried off by fever and starvation ; and finally that the miserable survivors had, after untold sufferings, at length emerged, with a few waggons and oxen, upon the higher and healthier country of the eastern borders of Damara and Ovampo-lands. After some delay, the Cape Government, prompted by feelings of humanity, despatched a ship with provisions and necessaries to Walfisch Bay, on the south-west coast, whence the much-needed aid was sent through Damaraland to the Boers.

Nothing more, whether of good or evil, was for some time heard by the outer world of the expedition ; but it was understood that the survivors had, to some extent, recovered themselves, and were gaining a livelihood by trading and hunting on the borders of Ovampoland. Early in 1887, tidings came that a Mr. W. W. Jordaan, a Cape colonist, had proceeded to Ovampoland in the year preceding, and had obtained a large concession of land from Khambonde, the young chief of the Omandonga Ovampos, that he had joined hands with the Boers, and had established a settlement under the name of Upingtonia—so called after Sir T. Upington, the late Premier of the Cape Government. It was further stated that the prospects of the new State were most promising. The ceded tract lay some 200 miles or so north-west of Walfisch Bay, between the eighteenth and twentieth degrees of south latitude, and included in its area the rich copper mines of Otavi, which for ages had been worked by the Omandongas.

It transpired that a large portion of the land was very fertile and produced bountifully maize, millet, or Kaffir corn, fruit, and vegetables, while the remainder was eminently adapted for pastoral purposes, and contained numerous strong fountains — a great desideratum in parched South Africa. The district was but sparsely inhabited by a few Berg Damaras and Bushmen, who paid tribute of ivory, copper ore, and ostrich feathers to the Ovampos. It may be here stated that Ovampoland is an undoubtedly rich country, lying immediately north of Damaraland, which, in its turn, is to the north of Great Namaqualand. The natives are a fine race of men, nearly allied to the true negro type of Central Africa. Armed with this rich concession, which he did not disguise he had obtained for a very inadequate consideration, Jordaan set about attracting colonists, and with such success that in November of 1886, there were settled some thirty or forty families of English, Germans, and Dutch Afrikanders, while other Boer families were also trekking in. Each family of settlers had allotted to it a farm of 3,000 morgen (6,000 acres). Shortly after, a bestuur, or council of thirteen members, was formed, on Boer lines, to administer the affairs of Upingtonia.

In the third act of this curious drama, a Mr. Lewis appeared on the scene, who, as principal adviser of Kamahero, chief of the Herero-Damaras, repudiated Khambonde's concession, alleging that the country ceded belonged to the Damaras, and in default of certain arrangements being made, he threatened to declare war upon both Upingtonians and Ovampos. In support of his claims, he urged that the Herero-Damaras had for centuries been possessed of the Otavi district; he denied that the Ovampos had ever worked the copper mines, but affirmed that his clients had worked them, and had sold the ore to the Ovampos, in exchange for knives, tobacco, beads, and other articles. In further proof, he asserted that there were no Ovampo names in that district, but that all local names were of Damara and Bushman origin, and that he could call hundreds of old inhabitants to prove his statements.

In the meantime, another cloud arose upon the horizon of the new settlement. Khambonde declared that when he executed the concession to Jordaan, he had merely intended to assign to him, in return for the guns and ammunition he received, a piece of land situated near his own kraal; that he had been tricked by Jordaan into the vast concession of the Otavi district, and that he would absolutely contest the claims of the bestuur of the new settlement. Further, in reply to a letter from the Damara chief, brought to him by Lewis, he despatched the following letter (presumably written for him by a missionary or trader):

"My dear eldest brother Kamahero,—Your letter I have received, and I must inform you that I have not sold the Otavi mine and the ground in question to Mr. Jordaan, and I have no idea whatever about the same being sold. All that I know is that I have sold to Mr. Jordaan a piece of ground near my station for guns and ammunition. So, as you are my eldest brother and adviser, please write and let me know what I must do with these things—whether I shall return them or not. I am only a child, and do not know the traditions of my forefathers, or even how far my boundary extends. So please, my eldest brother, kindly inform me how far my boundary extends to."

For a short time the dark curtain of the far interior drops, and then we have the final and tragic climax of the drama. On the 29th June, 1887, late in the afternoon, Jordaan " outspanned " at Ondonga, whither he had trekked to endeavour to come to terms with Khambonde. Considering the excited state of the Ovampos, and their character for treachery and bloodshed, for they had a few years before slaughtered a Roman Catholic mission, his journey was a dangerous one; but, being a man of strong will and determined courage, Jordaan seems to have had no misgivings.

After night had fallen, the waggon was surrounded by Ovampos, who had received their orders from the chief, and early next morning they advanced to the outspan, and squatted by the camp fire. Presently their leader spoke to Jordaan's driver, who was preparing breakfast, and told him that Khambonde had sent a present of an ox, and that he wished to salute him. Accordingly Jordaan was awakened, descended from his waggon, and greeted the Ovampos. He then prepared for breakfast, and just as he stretched across to take a cup of tea from his driver, one of the Ovampos fired both barrels of a heavy elephant gun into his chest at three paces, killing him dead on the spot. The Hottentot driver rushed to the waggon for a gun, but was forthwith despatched. The waggon was then taken to the chief and looted, while, shortly after, the body of the unfortunate Jordaan was buried in a hole scratched a foot deep in the earth. Afterwards the resident missionary erected a wall round the spot, and re-interred the remains. Two Boers and a trader in the neighbourhood, whose deaths had been also planned, had warning in time, and escaped with their lives. After this event, all was confusion in the settlement.

Some of the colonists at once trekked out, fearing greater disasters. The Boers first sent in hot haste to Damaraland to ask for German protection, but chiefly owing to difficulties of their own raising, after some negotiations, the German Acting Commissioner threw up the affair in disgust and left them to their own devices. But to add to their troubles, already sufficiently heavy, other entanglements ensued. A dispute had occurred between the Bushmen of the district and a wandering trader; the latter had taken the law into his own hands and shot one of the Bushmen, who in turn had retaliated by plundering the white man of his stores and looting his waggon. Both parties came before the bestuur of the settlement for justice, and as the Boers unduly favoured the trader, the Bushmen, who, to speak truth, had thus far behaved fairly well, after their imperfect lights, towards the settlers, took high offence, and departed, vowing that for the future they would follow their own laws and would make the Upingtonians smart for their shuffling. Meanwhile the settlers entered into fresh negotiations with the Damaras, and for a time it seemed that peace would yet prevail, and that by timely concessions to Kamahero, the Damara chief, they might even yet remain in possession of their new lands. A punitory expedition against the Ovampos to revenge Jordaan's murder was even talked of. But ill fortune was yet to cling to this ill-devised Colony. About this time there came news that Rudolph du Toit, field-cornet (sub-magistrate) of the new settlement, who had taken up a farm near Otavi mountain, had been foully slain by the revengeful Bushmen. Du Toit was a famous hunter, and hitherto from his knowledge of their ways and their language, had had considerable influence among the Bushmen, and had often employed them on his hunting expeditions. Trusting in these people, he had taken a number of them with him on a shooting trip as spoorers and hunters; but now, full of revenge and hatred against the Upingtonians, the Bushmen when out in the hunting veldt fell upon the Dutch hunter and killed him, and returning to his house ransacked the place. His wife and children, who were intended to be enslaved, escaped at night, and by great good fortune happened upon a party of traders and hunters. A commando was then quickly raised among the settlers, and the Bushmen, pursued to their mountain places, were surrounded and attacked. A hot fight, begun at grey of dawn, took place, in which the Bushmen, about fifty in number, lost some dozen in killed, while the Boers, only about twenty strong, were weakened by two deaths and the wounding of one or two others. The fight had straggled from a deep ravine into dense bush, and here the Bushmen, in their own peculiar element, had all the best of it, and at length the Boers having but half completed their campaign gave up the attack and returned to the main settlement.

In a very short time, the outlook became blacker than ever; the Damaras after all could not be appeased, the Ovampos were dangerously excited and gathering force for an attack, and the Bushmen, formerly half friends, were now bitter and exasperated enemies. Upingtonia, the intended stronghold and palladium of Dutch Afrikander freedom in South Africa, was now represented by a handful of Boer families—survivors of the long trek of 1876—of whom but eighteen were men capable of bearing rifles. Gathering their possessions and flocks and herds once more about them, this poor remnant inspanned their oxen, and again taking to their waggons, turned their faces for the Transvaal. Some of them passed through Damaraland and Namaqualand, and thence by the Orange River, after long months of incessant hardship, found their way back to their own land. Others took a different course, and, after great sufferings, won their way north of Lake N'Gami through the thirst-land, and thence across Bechuanaland back to Transvaal territory. A small band of the original trek remained behind at Humpata, near Mossamedes, in Portuguese territory, where they yet linger. Such was the miserable end of Upingtonia, a state founded too much upon the usual model of Boer filibusters. The Ovampo chief seems unquestionably to have been tricked into a huge concession, for nothing at all approaching an adequate consideration. A transaction thus inaugurated in duplicity, as it seems to have been, and terminated in bloodshed, could have only one certain result. The Ovampos, always suspicious of strangers, and disgusted with the dealings of the white men, will in future only yield their territory to force of arms, and will have no further intercourse if they can help it with Boers, British, or Germans. The result of this episode is that an important trade route is now practically closed for years to come.

Out of the great " Promised Land " expedition of years before, which had set forth with such high hopes, such boundless possibilities of land grabbing, such dreams of pastoral wealth, but a small and enfeebled remnant finally arrived back at Marico in the Transvaal. But though in one sense theirs had been a failure complete and miserable, in another sense the survivors had some reason for pride and gratulation. The expedition had endured for eleven long years, and in the annals of this nation of trekkers it will ever occupy the foremost place. In the earlier days of the Transvaal and Free State, nay, even of the Old Colony, the Boers became remarkable for the length and severity of their wanderings. Whole families existed for months and years in their waggons, moving slowly onwards with their flocks and herds, and such rude household belongings as they could carry with them. But there is no record at all approaching that of these Trek Boers, who had thus regained the Transvaal after eleven years of travel in the most dangerous and remote portions of South Central Africa. Even the Israelites in their wanderings never trekked more persistently or more doggedly than these Afrikanders. The foundation of the so-called Republic of Upingtonia was based upon foundations unwise, insecure, and improper, and its ending was not unjustly doomed to misfortune ; but the struggles of these frontier farmers had been of unexampled, nay, of incredible severity, and it is impossible to deny to the survivors some meed of admiration for their stubborn and undaunted pluck.

Chapter XX.

JAN PRINSLOO'S KLOOF.

A Legend of Cape Colony.

FAR away in the gloomiest recesses of a range lying between Zwart Ruggens and the Zwartberg, not far from where the mountains of that wild and secluded district give place to the eastern limits of the plateau of the Great Karroo, there lies hidden, and almost unknown, a kloof or gorge, whose dark and forbidding aspect, united to the wild and horrid legend with which it is invested, prevents any but the chance hunter or wandering traveller from ever invading its fastnesses. This kloof is about seven miles from the rough track that in these regions is dignified by the name of road; it is approached by a poort or pass through the mountains, and the way is, even for South Africa, a rough and dangerous one, although there are indications that a rude waggon track did formerly exist there. Standing upon the steep side of this kloof are the remains of what must have once been a roomy and substantial Boer farmhouse; but the four walls are roofless, the windows and doorways naked and destitute of sashes, the euphorbia, the prickly pear, and clambering weeds grow within and without, the lizard and snake abide there, and the whole appearance of the place denotes that many years have elapsed since Prinsloo's Kloof was tenanted by human life.

In many respects the wild kloof gives evidence that the Boer who first tarried there had an eye for good pasturage for his flocks and herds. The spekboom, and many another succulent bush dear to the goat breeder, flourish amid the broken and chaotic rocks with which the hillsides are strewn. A strong fountain of water runs with limpid current from the mountain at the back of the house; the flat tops of the hills around are clothed with long waving grasses, and the valley is eminently fitted to be the nursery of a horse-breeding establishment. A tributary of the Gamtoos River flows deeply, if fitfully, below the sheer and overhanging cliffs in a chain of pools called zee-koe gats (sea-cow or hippopotamus deeps)—the hippopotamus, though his name lingers behind, no longer revels in the flood—and the bottom of the valley is in many parts fertile and suited for the growth of grain and fodder crops.

Broken and uncouth as are many portions of the Witteberg and Zwartberg, the neighbourhood of Prinsloo's Kloof far surpasses them. There the volcanic action of a bygone age has perpetrated the most extraordinary freaks. The mountains are torn into shapes so wild and fantastic, that, viewed in profile against the red glow of the setting sun, all manner of weird objects may be conjured before the imagination. In some places, as the kloof runs into the heart of the hills, the cliffs' sides are so deep, so precipitous, and so narrow, that but little sunlight can penetrate beneath, and even on a hot day of African summer a chill strikes upon the spectator passing through.

It is not difficult to understand, from a Boer point of view, that this stern valley was a well-chosen spot in which to build a farmhouse. The distance from a roadway is, in Boer eyes, of no great account, and, as a rule, the farther from human habitation the Dutch farmer can get the better he is pleased. As for the forbidding aspect of the kloof, the stolid, unimaginative Boer would be little troubled on that score; for he has no eye whatever for picturesque or scenic effect, and will plant himself as readily upon the treeless wastes of the Orange Free State, or the most stony, barren mountain-side of the Old Colony, as in the most beautifully wooded country that South Africa can give him.

When Jan Prinsloo trekked into the kloof towards the end of the last century, the place must have been a very paradise and nursery of game. In the river the hippopotamus played, elephants roamed through the valleys and poorts everywhere around, the zebras ran in large troops upon the mountain tops, and many of the larger game, such as the koodoo, the buffalo, and the hartebeest, wandered fearlessly and free; while of the smaller game, such as rhebok, duyker-bok, and klipspringer, judging from the abundance of the present day, there must have been literally multitudes. To Jan Prinsloo, then, wild and sombre as the place was, it must have appeared, as he trekked down the pass, a veritable Boer elysium. But Jan, having played his part in the world—a part more fierce and turbulent even than was usual to the marauding frontier Boers of ninety years ago—made his exit from the scene in a manner cruel and horrible enough to match fitly with the rest of his wicked and violent existence.

Since Jan Prinsloo's fearful ending, which will be hereafter alluded to, the kloof has borne an evil reputation. Now and again a Boer has taken the farm, tempted by its pastoral advantages and its low purchase-money, but somehow, none have ever stayed upon it for long. The last tenant, an Englishman, quitted it hastily nearly thirty years ago, and ever since then the house has steadily fallen into ruin, the dark kloof has become year by year more sombre and more desolate, the footsteps of human beings now rarely penetrate thither, and even the very Kaffirs avoid the place.

In September of the year 1860, a young English Afrikander, Stephen Goodrick by name, who had, from the time he could handle a rifle, been engaged in the far interior, in the then lucrative, if dangerous occupation of elephant hunting; having amassed, at the age of thirty, some four or five thousand pounds, after fourteen years of hunting and trading in Northern Bechuanaland, and the Lake N'Gami region, threw up the game, and trekked down to Graham's Town with his last loads of ivory. These disposed of, and his affairs settled, he took unto himself for a wife, a handsome, dark-eyed girl, the daughter of Scotch parents, living near his own family in the Western province, and then set about looking for a farm, having determined to settle down to the more peaceful pursuits of pastoral farming. After a month of riding hither and thither, inspecting farms in the districts of Swellendam, Oudtshoorn, and George, none of which pleased his fancy, he turned his attention to the Eastern province.

Goodrick had been long and continuously away from the Colony, and in the brief intervals when he had rested from his hunting and trading expeditions, he had usually stayed with his father, an old colonist, in Swellendam, a district to the south-west of the Colony. His knowledge, therefore, of the Eastern province was necessarily somewhat restricted. Stephen, by chance, heard one day from a Boer trekking by with fruit and tobacco, that another Boer named Van der Meulen was leaving his farm near the end of Zwartberg. Losing no time, Stephen saddled up, bade temporary farewell to his wife, whom he left at his father's house, and traversing Lange Kloof and crossing the Kougaberg, he entered, on the afternoon of the third day, Prinsloo's Kloof, whither he had been directed.

It was a glorious hot afternoon in early summer, the sun shone as only it can in Africa, and under its brilliant rays and with the wealth of vegetation and flower life springing up everywhere around, the kloof, savage though it appeared, put on its mellowest aspect; and as Goodrick rode up to the farmhouse and noticed the flocks and herds, all sleek and in good condition, he thought that there might be many worse places to outspan for life, than in this beautiful, if solemn valley.

At the farmhouse he was welcomed by the owner, Van der Meulen, and after a stroll round the kraals and supper over, a business conversation took place before the family retired to rest, which, as it seemed to the young Englishman, they did hurriedly and with some odd glances at one another. Next morning all were up early, and Goodrick rode round the farm—all good mountain pasture, embracing some 19,000 morgen (rather more than 40,000 acres) in its area. The Boer, in his uncouth, rough way, warmly praised the farm; the price he asked was extremely small, and the annual Government quit rent very trifling. Van der Meulen explained as his reason for selling the place apparently so much below its value, that he had been offered, at an absurdly small price, a very fine farm in the Transvaal, by a relation who had lately annexed the best of the land of a native chief; and as many of his blood relations, Boertrekkers of 1836, were settled there, he wished to quit the Colony quickly and join them. Finally, Goodrick agreed to buy the farm, together with part of the stock, and early on the following morning left the kloof. The purchase was shortly completed at Cape Town, where the vendor and purchaser met a week afterwards, and the Van der Meulens having trekked out with all their household goods and belongings, the Englishman and his wife prepared to enter upon their property.

Stephen Goodrick, then, with two waggons, carrying his wife, her white female servant, a quantity of furniture and household and farming necessaries, and taking with him four Hottentots and half-a-dozen horses, trekked again through Lange Kloof, over the Kougaberg, and thence through a country partly mountain, partly karroo, until one afternoon early in October, the waggons crossed the deep and dangerous drift of the river, and went up through the poort that led into Prinsloo's Kloof. After a most difficult and tedious piece of travelling for some seven miles—for the half-forgotten waggon track lay up and down precipitous ascents and declivities, littered here and there with huge boulders, or hollowed out into dangerous spruits and holes—at length the stout but wearied oxen faced the last steep hill to the farmhouse, and with many a pistol crack of the great whip, many a Hottentot curse directed at Zwartland, Kleinboy, Englesman, Akerman, and the rest, dragged their heavy burdens up to the open space that had been cleared in front of the homestead. It had been arranged that Van der Meulen's eldest son should remain upon the farm until Goodrick and his wife had arrived, and, further, that an old Hottentot, Cupido by name, who knew the farm and its ways well, and two young Kaffirs who had lately arrived from Kaffraria in search of work, should transfer their services to the new-comer.

These four were therefore ready, having already brought in and kraaled the goats for the night, and they assisted the Englishman to outspan his oxen and unload the waggons. After two or three hours' hard work, a good portion of the waggons was unloaded, and part of the furniture arranged in the house; three of the horses were placed for the night in the rough building adjoining the dwelling-house, that served for a stable, while the remainder had been turned into a large stone kraal which lay on the other flank of the house. Meanwhile the white servant had prepared supper, which partaken of, the wearied travellers retired to rest. About the middle of the night Goodrick and his wife were suddenly aroused by a great commotion in the stable; the horses were trampling, plunging, and squealing as if suddenly disturbed or scared. Then there rose upon the night, as it seemed just outside the house, a wild scream, hideous in its intensity and full of horror.

Hastily thrusting on some clothes and taking a lantern, Goodrick ran round to the stable. The night, though there was no moon, was not dark, and the stars shone clear in the firmament above. Nothing was to be seen, no sound could be heard save the snorting of the horses, and the weird cry of a leopard, (strangely different, as the hearer well knew, from the scream heard just previously), that sounded from the rocks a mile or so away on the right. Quickly entering the stable, Stephen was astonished to find the horses in a profuse sweat, trembling, their halters broken, their eyes startled and excited, and their whole demeanour indicating intense fear. What could be the cause? There was, apparently, no wild animal about, nothing in the stable calculated to excite alarm; the animals were old comrades, and not likely to have been fighting. Goodrick was altogether puzzled, and, leaving the stable, went to a shed in rear of the house, where the natives slept, and roused the old Hottentot. He could give no reason for the disturbance. "Wolves (hyænas) were not likely to approach the house, and the tigers (leopards) had not been very troublesome lately," and he could think of nothing else to explain the matter. There was a scared look in the old man's face, which Goodrick thought nothing of at the time, but which he afterwards remembered. After some little trouble, fresh halters were procured, the horses tied up and soothed, and the two again retired, Cupido being cautioned to keep his ears open against further disturbance; nothing further occurred during the night.

Next morning, after seeing the goats unkraaled, watered, and despatched to their day's pasturage in charge of the two Kaffir herds, Goodrick asked the young Boer at breakfast if he had heard the noise among the horses, and the wild scream, and what could be the cause, and if there were any cattle stealing about this wild neighbourhood. Young Van der Meulen's heavy, immovable countenance changed slightly, but he replied "that he could give no explanation, except that perhaps a leopard might have been prowling about; they were pretty numerous in the kloof." Stephen explained that he and the Hottentots had spoored everywhere around the stable for leopards, but could find no trace.

Here the subject dropped, and Van der Meulen relapsed into silence, except when the Englishman asked him what game there was about the hills. "You will find," he said, "plenty of small buck—klipspringers, rhebok, and duykerbok, and there are still a fair number of koodoo, which take some stalking though. Then on the berg tops there are several troops of zebras, as well as hyænas and leopards; but the zebras we have seldom shot, they take so much climbing after, and you know we Dutchmen prefer riding to walking. You will find still lots of springbok and steinbok, and some black wildebeest (gnu) on the plains beyond the mountains. Yes, I have had many a 'mooi schiet op de plaats' (pretty shoot on the farm)." Suddenly the young man's heavy features changed again as he said, "Allemaghte! (Almighty), but I shall be glad to get out of this place; I hate it! I want again to get on to the Transvaal high veldt, where I trekked through two years ago, and where you can shoot as many blauuw wildebeest (brindled gnu), blessbok, quagga, springbok, and hartebeest, as you want in a day's ride. Ja! that is the land for me; these gloomy poorts and kloofs are only fit for leopards and 'spooks' (ghosts). Then, you know, Mynheer, the Transvaal is free; we never loved your Government, which is always wanting from us this, that, and the other, and I shall be glad to trek out. Up in Zoutpansberg we shall be able to hunt the kameel (giraffe), and the zwart wit pens (sable antelope), and elephant, as much as we like, and for our winter pasture we shall not have to pay a single rix-dollar. Ja! I have had enough of Prinsloo's Kloof, and never wish to see it again." This long speech delivered, the Boer relapsed into silence. There was a curious look on the young man's face, as he had spoken, which Goodrick and his wife could not quite define or understand.

An hour afterwards Van der Meulen had slung his rifle on his back, packed some biltong (sun-dried meat) in his pockets, saddled up his horse, and bidden farewell to the tenants of the kloof. The Englishman and his young wife watched his retreating form as it slowly proceeded down the valley, and presently disappeared amidst a grove of acacia trees that margined the river; then they turned to the house. "I don't quite understand that fellow," said Stephen, "do you, Mary? I can't help thinking there was something behind what he said. Why were his people so eager to leave this farm? However, darling, the farm *is* a good one and a cheap one, we are young and strong, and ought to be as happy as any Afrikanders in the Colony."

"Yes, Stephen," said his wife, "I thought there was something queer in what the young man said, but it could have been only fancy. I am sure we ought to be happy and contented, and with you by my side I shall always be so."

In a few weeks time Goodrick had increased his stock of goats, and had bought a sufficient number of horses to start a stud farm upon the mountains around. Things seemed to be going well with him. The pasture was in splendid condition, the valleys and kloofs that led into the mountains literally blazed with flowers of every conceivable hue, from the great pink or crimson blossomed aloes, that gave warmth to the towering brown rocks above, to the lovely heaths, irises, and geraniums that clothed as with a brilliant carpet the bottom grounds. The house had been thoroughly cleansed, put into order, and the new furniture settled into it, and young Mrs. Goodrick busily employed her days in household duties. Her husband had had several good days shooting about the hills, and had brought in two koodoos (one of the largest and most magnificent of South African antelopes), whose noble heads and heavy spiral horns now adorned the dining-room, besides many a head of smaller antelopes and innumerable francolins, pheasants, ducks, and other feathered game.

Yet somehow, though things had so far gone well, the young couple were not quite comfortable. The disturbances among the horses, although not repeated for several nights, had occasionally happened; the same horrid scream had been heard, and their causes had so far completely baffled Stephen Goodrick. He had tried all sorts of plans, changed the horses, and even had them all turned loose together in the great stone kraal, but with the same results. They were found over and over again at night, mad with fear and drenched with sweat, trampling and plunging in the stable, or tearing about the enclosure.

Cupido and the Kaffirs, and his own Swellendam Hottentots, had been questioned and cross-examined, but to no purpose. Twice had Goodrick remained on the watch all night. On one occasion he believed he had seen a figure move quickly past him in the dark night, and the horses had been disturbed at the same time; but nothing further could be traced, and no spoor of man or quadruped was ever discovered. The thing was a mystery. At length, one moonlight night, Goodrick ran out, hearing the now familiar noises, and taking with him his great brindled dog "Tao" (so called from the Bechuana name for lion, bestowed upon him when, as a puppy, his master had first taken him up into the interior, and he had, on his introduction to the king of beasts, single-handed, driven his tawny majesty from his shelter in a dense reed-bed), who had often hunted elephant, rhinoceros, buffalo, and lion, he quickly went round to the stable.

At this moment the two Kaffir herds also came running out on hearing the noise. Just as they approached the stable together they beheld a figure pass through the open doorway, as they supposed, and swiftly glide away to the hillside. The dark figure was clad in a broad-brimmed Boer hat and quaintly-cut old-fashioned looking dress, as Goodrick could plainly notice. Stephen shouted, and with the Kaffirs gave chase, but after a few minutes' running the man suddenly vanished into the bushy scrub that grew on the mountain-side, and no further trace could be found, although the Kaffirs hunted everywhere around.

Meanwhile, Stephen turned for his dog, surprised that the animal, usually so fierce and impetuous, had not led the chase. To his utter astonishment, "Tao" was close at his heels, his tail between his legs, his hackles up, and with every symptom of terror upon him. The thing was incomprehensible; the dog had never feared man or beast in his life before; and many a time and oft had faced, as they turned at bay, the fierce and snarling lion, the dangerous sable antelope with his scimitar-like horns, and the wounded and screaming elephant. At length, turning back, they entered the stable; to their surprise the door was locked, and on being opened the horses as usual were loose, and in the last extremity of fright. Nothing more could be done that night. In the morning the Kaffirs and Hottentots searched everywhere for spoor, but could find no trace of the midnight marauder. Cupido, indeed, shook his head, rolled his bloodshot-looking eyes, and appeared to take the ocurrence as a matter of course.

Three mornings afterwards the two Kaffirs came to Stephen, declared that they had seen on the previous night the same dark figure just outside their sleeping shed; that the terrible appearance of the face of this apparition, which they saw distinctly in the moonlight, had made them utterly sick and terror stricken; that the thing was a thing of witchcraft, and that nothing would induce them to stay another night on the farm.

"We like the N'kose (chief or master) well," they said, "but we dare not stay longer in this country, or we shall be slain by the witchcraft we see around us; why do you not get a 'smeller out' to cleanse this place from the evil?" The two men, who in daylight were, as most Kaffirs are, bold, hardy fellows, were evidently in earnest in what they said, and though Goodrick, who could ill afford to part with them at a moment's notice, offered them increased wages, they steadfastly declined; and at length finding he could not shake their resolution, he reluctantly paid them their money and let them go. Goodrick learned some months afterwards from a friend, that these men had marched straight for the boundary of the Colony, crossed the Kei, and rejoined their own tribe, the Gaikas, in Kaffraria.

Goodrick now began to think somewhat seriously of the matter, and to ask himself with inward misgivings what it meant. Brave man though he was, like most mortals he was not quite proof against superstition, and he began to find himself half fearing that there was something not quite canny about the place. How else could he account for the locked door, the suddenly vanishing figure, the sickening yell, and the lack of footmarks? However, he kept his thoughts from his wife, and made some excuse about a quarrel with the Kaffirs as to wages, to explain their sudden departure. She, although accepting the explanation, seemed uneasy, and at last burst out, "Oh, Stephen, I think there is something wrong about this kloof—

some dreadful mystery we know nothing of. Have you ever noticed that even the Kaffirs in the kraal a few miles away beyond the poort never enter here? Not a soul amongst the farmers comes near us, and as for 'Tao,' he never seems happy now, and is always restless, suspicious and alarmed."

That same night the wild, unearthly scream rose again; the same tumult was heard in the stable, and Stephen rushed out; and once again he saw the figure passing in front of him. This time he had his rifle loaded, and after calling once, fired. Still the figure retreated; another shot was fired, but to no purpose; the figure apparently glided imperceptibly onwards, and then suddenly disappeared, as it seemed, sheer into the earth. Goodrick knew so well his powers with the rifle, with which he had been famous as a deadly shot, that he could not bring himself to believe he had missed twice within fifty yards. From this incident he could form no other conclusion, and he shivered as he thought so, that the night disturber was not of human mould.

Meanwhile the horses were becoming worn to shadows, their coats stared, they lost flesh and looked altogether miserable. Fresh horses had been brought in, but the effect was ever the same. Shortly after, two of the Swellendam Hottentots left, and the other two, with Cupido and Mrs. Goodrick's servant, alone remained. Goodrick was now in great straits; he could not immediately procure other native servants, and only managed to get through his farm work with the greatest trouble and labour.

Things drifted on uncomfortably for another week or two, and each day, as it came and went, seemed to Goodrick and his wife to increase the gloom and uncertainty of their life in the kloof. At length a climax arrived. Christmas, but a sombre one, had sped, and South African summer, with its heat, its flies, and other manifold troubles, was now at its height.

On the 15th of January, 1861, a day of intense heat was experienced. All day the landscape had sweltered under a still oppression that was almost unbearable, and the very animals about the farm seemed touched and depressed by some mysterious influence.

Towards nightfall dark clouds gathered together suddenly in dense masses; in the distance, long, rolling thunder-peals were heard approaching in strangely slow, yet none the less certain movement. Cupido, the old Hottentot, had fidgetted about the house a good deal all the evening, and finally, just before ten o'clock, he asked his master if he might for that night sleep on the floor of the kitchen, in order, as he put it, to attend more quickly to the horses, if anything scared them. Goodrick noticed that the old man looked agitated, and good-naturedly said " Yes."

Still slowly onward marched the stormy batteries of the sky, until at eleven o'clock they burst overhead with a terrific crash (preceded by such lightning as only Africa can show), that literally seemed to tear and rend each nook and corner of the gorge, reverberating with deafening repetition from every krantz and hollow and rocky inequality in the rude landscape. Rain fell in torrents for a time, then ceased. Again and again the thunder broke overhead, while the lightning played with fiery tongue upon mountain and valley, showing momentarily, with photograph-like clearness, every object around. Sleep on such a night was out of the question, and Goodrick and his wife sat together listening to the hideous tumult with solemn faces. At length, at about twelve o'clock, the storm for a brief space rolled away, only to return in half an hour with increased severity.

Goodrick had gone for a few moments to the back door, which faced partly towards the entrance to the kloof, and found Cupido standing there seemingly listening intently. As the tempest approached again with renewed ferocity, some strange confused noises, shrieks and shouts as it seemed, were borne upon the strong breeze that now preceded and hurried along the thunder clouds.

"Hallo!" said Goodrick, "what the deuce is that? There surely can't be a soul about on such a night as this?" Again a hideous scream was borne up the valley. "Good God! that's the very yell we've heard so often round here at night," repeated the Englishman. "It's not leopard, it's not hyæna; what on earth is it, Cupido?" The Hottentot was now trembling in every limb; his yellow, monkey-like face had turned ashy grey, and his bleared eyes seemed full of some intense terror. "Baas (master), it's Jan Prinsloo's night, and if you're wise you'll shut the doors fast, pull down the blinds, and not stir or look out for an hour."

"What do you mean, man?"

"I mean that the ghosts of Jan Prinsloo, who was slain here years ago, and his murderers, are coming up the kloof." At that moment the cries and shoutings sounded up the valley closer and closer, and it seemed as if the rattling of horses galloping along the rock-strewn path could be distinguished through the storm. Just then the other two Hottentots, who at length had also heard the din, rushed across from their shed, and huddled into the kitchen. Mrs. Goodrick at the same instant ran into the room. "What is the matter, Stephen?" she cried; "I am certain there is some dreadful work going on." "Yes, wife, there is some devilish thing happening, and I mean to get to the bottom of it. I haven't hunted fifteen years in the interior, to be frightened by a few strange noises." So speaking, the young farmer went to the sitting-room, took down and quickly loaded two rifles and his revolver, and returned to the kitchen. Handing one rifle to the Hottentot, he said, "Here, Cupido, take this; I know you can shoot straight, and, if needful, you'll have to do so. Wife, give the Totties a soupje each of brandy."

This was quickly done, and the result seemed, on the whole, satisfactory, and the Hottentots more reassured. In a few more seconds the storm burst again in one appalling roar, and after it could now be heard the clattering of hoofs up the hillside, mingled with shrieks and shouts. This time the tempest passed rapidly overhead, the dense black clouds rushed on, and suddenly the moon shone out with wonderful brightness.

Onward came the strange noises, sweeping past the side of the house, as if up to the great stone cattle kraal, that lay sixty yards away. Then was heard the loud report of a gun. Stephen could stand it no longer. "Come on, you fellows, with me," he exclaimed, as he ran out towards the kraal. Cupido and Mrs. Goodrick, who would not be left behind, alone followed him; the white servant woman, and the remaining two Hottentots stayed in the kitchen, half dead with fright, the one on a chair, her apron clasped to her head and ears, the others huddled up in a corner. The three adventurers were not long in reaching the kraal, whence they heard the same dreadful cries and shrieks proceeding, mingled with the trampling of feet. Goodrick first approached the entrance, which he found burst open. The sight that met his eyes, and those of his wife and Cupido, close behind, was enough to have shaken the stoutest heart.

Under the clear illumination of the moon, which now shone forth calm and serene, the enclosure seemed as light as day. In the far corner, to the right-hand, seventy paces distant, the half-dozen horses, that had been turned in, stood huddled with their heads together like a flock of sheep. On the opposite side, from the entrance, a frightful-looking group was tearing madly round. First ran a tall, stout figure, clad in the broad-brimmed hat and quaint old-fashioned leathern costume, which Goodrick in a moment recognised. In its hands it grasped a huge, long, old flint "roer," a smooth-bore elephant gun, such as the Boers used in earlier days. The figure, as it fled, had its face half-turned to its pursuers, who consisted of six half-naked Hottentots armed with assegais and knives. As the chase, for such it was, swept round the kraal and the figures approached the entrance, every face could be plainly discerned; and this was the horrible part of it. These faces were all the faces of the dead, gaunt, ghastly, and grim, and yet possessed of such fiendish and dreadful expressions of anger, cruelty, and lust for blood, as to strike a chilling terror to the hearts of the three spectators. Brave man and ready, though he was, Goodrick felt instinctively that he was in the presence of the dead, and his rifle hung listlessly in his hand.

Closer the fearful things approached the spellbound trio, till, when within thirty yards, the leading figure stumbled and fell. In an instant, with diabolical screams, the ghostly Hottentots fell upon their quarry, plying assegai and knife. Again the awful scream that the kloof knew so well rang out upon the night; then followed a torrent of Dutch oaths and imprecations; and then the dying figure, casting off for a moment its slayers, stood up and laid about it with the heavy "roer" grasped at the end of the barrel.

The three living beings who looked upon that face will never to their dying days forget it. If the expression of every crime and evil passion could be depicted upon the face of the dead, they shone clear under the pale moonlight upon the face of the dying Dutchman—dying again though dead. Once again with wild yells, the Hottentots closed on their victim, and once more rang the fiendish dying yell. Then, still more awful, the Hottentots, as it seemed in an instant, stripped the half dead body, hacked off the head and limbs, and tore open the vitals, with which they bedabbled and smeared themselves as they again tore shrieking round the kraal. Flesh and blood could stand the sight no longer; Mrs. Goodrick, who had clung to her husband spell-bound during the scene, which had taken in its enactment but a few seconds, fainted away. Goodrick turned to take his wife in his arms, with the intention of making hurriedly for the house. At that instant the horrid din ceased suddenly, and was succeeded by a deathly silence. Turning once more to the kraal gate, Goodrick at once perceived that the whole of the enactors of this awful drama had vanished. He rubbed his eyes in vain to see if they deceived him, but a nod from the half-dead Cupido convinced him that this was not so. No, there was no doubt about it, the waning moon cast her pure and silvery beams calmly and peacefully upon a silent scene. Not a trace of the bloody drama remained; not a whisper, save of the soft night breeze, told of the dreadful story.

"Baas," whispered the Hottentot, "they'll come no more to-night." Quickly Goodrick raised his fainting wife and carried her into the house, where, after long and anxious tending, she was restored to consciousness. Placing her in the sitting-room, upon a couch that he had himself made from the soft skins, "brayed" by the Kaffirs, of the antelopes he had shot, he at length induced her to sleep, promising not for a moment to leave her, and with his hands clasped in hers.

At length the night wore away, the sun of Africa shot his glorious rays upward from behind the rugged mountain walls of the kloof, and broad daylight again spread over the landscape. Goodrick was glad indeed to find that with the bright sunshine his wife, brave-hearted woman that she was, had shaken off much of the night's terrors; yet were her nerves much shaken. For the last time the goats were unkraaled, and sent out, with the two somewhat unwilling Hottentots, to pasture. Breakfast and some strong coffee that followed this operation made things look brighter; and then taking the couch, and setting it upon the stoep (verandah), just outside the windows of their room, and placing a chair for himself, Goodrick went out to the back and called Cupido in with him to the "stoep," where he made the little ancient yellow man squat down. "Cupido," said he, "I am going to inspan this morning, load up one of the waggons, and send my wife and servant under your charge out of this cursed place to Hemming's farm—the next one, twenty-five miles away, out on the karroo. To-morrow, with the help of some Kaffirs I shall borrow from Mr. Hemming, I shall get down the horses from the mountain, load up both the waggons with the rest of the furniture and farm tackle (as soon as you return, which you will do very early), and trek out of the kloof, never again to set foot in it. But first of all, you will tell me at once, without lying, why you have never said a word to me of this horrible secret, and what it all means. Now speak, and be careful."

"Well, baas," said Cupido, speaking in Boer Dutch, the habitual language of the Hottentots, "you have been a kind baas to me, and de vrouw" (nodding to his mistress) "has been good to me too; and I will tell you all I know about the story. I would have warned you long ago, but Mynheer Van der Meulen, when he left, made me promise under pain of being shot, and I believe he was Boer enough to have kept his word, for he often gave me the sjambok,* and I dare not. I was born here in the kloof many years ago, many years even

---

* The sjambok is a short tapering whip made of hippopotamus hide, and is a most cruel and punishing weapon in bad hands.

before slavery was abolished and the emigrant Boers trekked out into the Free State and Transvaal, and you will know that is long since.

"My father lived as a servant under that very Jan Prinsloo, whom you saw murdered last night in yonder kraal, and many a time has he told me of Prinsloo and his evil doings and his dreadful end. Well, Jan Prinsloo was a grown man years before the English came across the shining waters and took the country from the Dutch.* He was one of the wild and lawless gang settled about Bruintjes Hoogte, on the other side of Sunday River, who bade defiance to all laws and governments, and who, under Marthinus Prinsloo (a kinsman of Jan's) and Adriaan Van Jaarsveld, got up an insurrection two years after the English came, and captured Graaff Reinet.

"General Vandeleur soon put this rising down, and Marthinus Prinsloo and Van Jaarsveld were hanged, but Jan Prinsloo, who was implicated, somehow retired early in the insurrection, and was pardoned. Some years before this, Jan was fast friends, as a younger man, with Jan Bloem, who, as you may have heard, was a noted freebooter who fled from the Colony across the Orange River, raised a marauding band of Griquas and Korannas, and plundered, murdered, and devastated amongst many of the Bechuanas tribes, besides trading and shooting ivory as well. The bloody deeds of these men yet live in Bechuana story. Jan Bloem at last, however, drank from a poisoned fountain in the Bechuana country, and died like a hyæna, as he deserved. Then Jan Prinsloo took all his herds, waggons, ivory, and

---

* The Cape was first taken by the British in 1795.

flocks, came back over the Orange River, sold off the stock at Graaff Reinet, and came and settled in this kloof. He had brought with him some poor Bechuanas, and these people, who are in their way, as you know, great builders in stone, he made to build this house, and the great stone kraal out there, where we saw him last night. He had, too, a number of Hottentots, besides Mozambique slaves, and these he ill-treated in the most dreadful manner, far worse even than any Boer was known to, and that is saying much. At last one day, not long after the Bruintjes Hoogte affair, he came home in a great passion, and found that two of the Hottentots' wives and one child had gone off without leave to see some of their relatives, Hottentots, who were squatted some miles away.

"When these women came back in the evening, Prinsloo made their husbands tie them and the child to two trees, and then and there, after flogging them frightfully, he shot the poor creatures dead, child and all. As for the husbands, he 'sjambokked' them nearly to death for letting their wives go, and then turned in to his 'brandewein' and bed. That night all his Hottentots, including seven men who had witnessed the cruel deed — God knows such deeds were common enough in those wild days — fled through the darkness out of the kloof, and never stopped till they reached the thick bush-veldt country, between Sunday River and the Great Fish River. Just at that time, other Hottentots, roused by the evil deeds of the Boers, rose in arms, and joined hands with the Kaffirs, who were then advancing from beyond the Fish River.*

---

* This rising occurred in 1798-9.

"Well, the Kaffirs and Hottentots, to the number of 700, for some time had all their own way, and ravaged, plundered, burned, and murdered the Boers and their farms, even up to Zwartberg, and Lange Kloof, between here and the sea. While they were in that neighbourhood, a band of them, inspired by the seven Hottentots of Prinsloo's Kloof, came up the Gamtoos River, in this direction, and met with Jan Prinsloo and a few other Boers, who were trekking out of the disturbed district with their waggons, and who had come to reconnoitre in a poort, fifteen miles away from here. All the Boers were surprised and slain, excepting Prinsloo; and while the Kaffirs and other Hottentots stayed to plunder the waggons, Prinsloo's seven servants, who were all mounted on stolen horses, chased him like 'wilde honde'* hunting a hartebeest, for many hours; for Jan rode like a madman, and gave them the slip for three hours, while he lay hid up in a kloof; until, at last, as night came on, they pressed him into his own den here.

"It was yesterday, but years ago, just when the summer is hottest and the thunder comes on, and just in such a storm as last night's, the maddened Hottentots, thirsting for the murderer's blood, hunted Prinsloo up through the poort. They were all light men and well mounted, and towards the end gained fast upon him, although Jan, who rode a great 'rooi schimmel' (red roan) horse, the best of his stud, rode as he had never ridden before. Up the kloof they clattered, the Hottentots close at his heels now. Prinsloo galloped to the great kraal there, jumped off his horse, and ran inside, like a leopard among his

---

* The wild hunting dogs of the Cape (*Lycaon pictus*).

rocks, fastening the gate behind him, and there determined to make a last desperate stand for it.

"The Hottentots soon forced the gate and swarmed over the walls, not however before one was killed by Prinsloo's great elephant 'roer.' Round the kraal they chased him, giving him no time to load again; at last, as you know, he fell and was slain, and the Hottentots cut off his head, and arms, and legs, and tore out his black heart, and in their mad, murderous joy and fury, smeared themselves in his blood. Then the men looted the house, set fire to what they could, and afterwards rejoined their comrades next morning. They told my father, who had known Prinsloo, the whole story when they got back. These six men were all killed in a fight soon afterwards when the insurrection was put down; and the Kaffirs and Hottentots were severely punished.

"Well, ever since that night the thing happens once a year upon the same night. Many Boers have tried to live in this place since that time, but have always left in a hurry after a few week's trial. I believe one man did stay for nearly two years; but he was deaf, and knew nothing of what was going on around, until one Prinsloo's night; when he saw something that quickly made him trek. I once saw the scene we witnessed last night; it was many years ago, when I was a young man in the service of a Boer, who had just come here; (before then I had been with my father in the service of another Boer, forty miles away towards Sunday River). Next morning after seeing Prinsloo and his murderers, my master trekked out terror stricken. I never thought to have seen the horrible thing again, but eight months ago when the Van der Meulens came here, I was hard up and out of work, and though I didn't half like coming into the kloof again, I thought, perhaps, after so many years, the ghosts might have vanished. I hadn't been many nights here, though, before I knew too well I was mistaken. Even then I would have left, but Van der Meulen swore I should not. He and his family came here soon after Prinsloo's night, and left before it came round again; but after the old man and his sons had twice been face to face with Jan's ghost prowling about the stable and kraals, and even looking in at the windows, they were not long before they wanted to clear out, and now you know their reason, baas."

"Yes, Cupido, to my cost I do," said Goodrick. "I don't suppose I shall ever come across that delightful family again; for it is a far cry to Zoutpansberg in the north of the Transvaal, and a wild enough country when you get there. But tell me, why is it that this dreadful thing is always in and out of the stables and kraals, frightening the horses?" "Well, baas, I am not certain, but I believe, for my father aways told me so, that Prinsloo was very fond of horseflesh, extraordinarily so for a Boer; for you know as a rule they don't waste much care on their horses, and use them but ill. He had the finest stud in the Colony, and took great pains and trouble with it; and they say that Jan's ghost is still just as fond as ever of his favourites, and is always in and out of the stable in consequence. Anyhow the horses don't care about it, as you know; they seem just as scared at him as any human being."

Cupido, like all Hottentots, could tell a story with the dramatic force and interest peculiar to his race, and the bald translation here given renders very scant justice to the grim legend that came from his lips. After the quaint little yellow man had finished, Mrs. Goodrick gave him some coffee, and immediately afterwards the party set about loading up one waggon with a part of the furniture; and this done, and Mrs. Goodrick and her servant safely installed, Cupido, the oxen being inspanned, took the leading reim of the two first oxen, and acted as leader, while Goodrick sat on the box and wielded the whip.

Twelve miles away beyond the poort that opened into the kloof there was a Kaffir kraal, and having arrived there, Goodrick was able to hire a leader; and Cupido having relieved his master of the whip, and received instructions to hasten to Hemming's farm as quickly as possible with his mistress, Goodrick saddled and bridled his horse, which had been tied to the back of the waggon, and rode back to his farm. The night passed quietly away; the two remaining Hottentots begged to be allowed to sleep in the kitchen, and this favour their master not unwillingly accorded to them. Next morning, at ten o'clock, Cupido, who had trekked through a good part of the night, arrived, and with him came Mr. Hemming, the farmer, and four of his Kaffirs. Hearing of his neighbour's trouble, and having seen Mrs. Goodrick comfortably settled with his own wife, he had good-naturedly come to his assistance. "So Jan Prinsloo has driven you out at last," said he, upon meeting Goodrick. "I heard from your wife last night what you had seen the night before. I was afraid it would happen, and would have warned you in time if I had known. But I never even heard that the Van der Meulens had sold the farm, till they had cleared out, and I met you about a month after you had been here; and as you were a determined looking Englishman, and the half-dozen people who have tried the farm in the last twenty years have been superstitious Dutch, I thought perhaps you might succeed in beating the ghost, where they failed. I haven't been in the kloof for many years, and after this experience, which bears out what my father and others who knew the story well have always told me, I shan't be in a hurry to come in here again. It is a strange thing, and I don't think, somehow, the curse that seems on the place will ever disappear." "Nor I," said Goodrick, "I'm not in a hurry to try it. I never believed in ghosts till the night before last; for I never thought they were partial to South Africa, but after what I saw I can never again doubt upon that subject. The shock to me was terrible enough, and what my wife suffered must have been far worse."

With the willing aid of his neighbour and his Kaffirs, as well as his own Hottentots, Goodrick got clear of the kloof that day, and after a few days spent at Mr. Hemmings', trekked away again for Swellendam, to his father's home. Six months later he finally settled in a fertile district not far from Swellendam, where he and his wife and family still remain. Cupido died in his service only a few years since. After much trouble Goodrick sold his interest in Prinsloo's Kloof and the farm around, for a sum much less even than what he gave Van der Meulen for it; and it is only fair to say he warned the purchaser of the evil reputation of the place before this was done. It is a singular fact that on his way to take possession of the kloof the new purchaser fell ill and died, and the place has never since been occupied.

Although it is nearly thirty years since these events took place, and Mrs. Goodrick has now a grown up family around her, she has never quite thrown off the terror of that awful night. Even now she will wake with a start if she hears any sudden cry in her sleep, thinking for the moment it is the death scream of Prinsloo's Kloof. As for the haunted kloof, it lies to this day in desolation black and utter. No footfall wakes its rugged echoes; the grim baboons keep watch and ward; the carrion aas-vogels * wheel and circle high above its cliffs, gazing down from their aerial dominion with ever-searching eyes; the black and white ravens seek in its fastnesses for their food, looking, as they swoop hither and thither, as if still in half mourning for the deed of blood of bygone years; and the antelopes and leopards wander free and undisturbed. But no sign of human life is there, or seems ever likely to be; and if by cruel fate, the straying traveller should haplessly outspan for his night's repose, by the haunted farmhouse, on the night of the 15th of January, he will yet see enacted, so the neighbouring farmers say, the horrible drama of Jan Prinsloo's death.

---

* Vultures.

Chapter XXI.

THE TRUE UNICORN.

THE Cape of Good Hope claims as the supporters of its coat of arms two antelopes—one, perhaps, the noblest, the other the most fantastic, of that astonishing array with which South Africa is so bountifully endowed. Of these, the famous gemsbok (pronounced by the Boers, gutturally, "hemsbok") is one; the curious black wildebeest, or white-tailed gnu (now, by the way, much scarcer than the brindled gnu), the other. Until towards the middle of this century both these antelopes were to be found upon every wide-spreading karroo of the Cape Colony. But, such has been the fierceness of the war of extermination waged against the nobler of the fauna, the black wildebeest (*Catoblepas gnu*) is now almost practically extinct within the ancient colonial limits—saving on one or two farms, where they are yet preserved, in the Western province; while the peerless gemsbok (*Oryx capensis*), the prototype of the fabled unicorn, has, with the exception of the small number that still linger in the wastes of Northern Bushmanland, been in the same manner exterminated.

A very noticeable feature in connection with the fauna of Southern Africa is the apparent fickleness of its geographical distribution. Many animals, for some unsolved reason, have their habitats peremptorily bounded by a river, or a belt of sand or jungle. The gemsbok, never profusely scattered, is now only to be found in the Kalahari desert—its chief stronghold—upon the fringes of Great Namaqualand and Damaraland, in Bechuanaland, where it only occurs on the Kalahari border, except to the northward, where it is occasionally found in the interior, and, very sparsely, in the north-west of Cape Colony. In the more jungly countries of the Eastern province of the Cape, of Kaffraria and South-Eastern Africa, and the Zambesi regions, although vast plains here and there occur, it is utterly unknown. Almost independent of water, it loves the parched and arid regions of South-Western Africa alone, and the dryer and more desolate the terrain the better does this antelope appear to appreciate it. The oryx and the springbok, indeed, seem created by nature to adorn and beautify these wastes. The gemsbok, from the rare singularity of its beauty, its consciously majestic carriage, and the difficulties of procuring it, may fairly be classed among the five noblest antelopes of the world. In that wondrous galaxy of South African antelopes, numbering more than twenty-five species in all, it rightly stands upon the same proud vantage ground as the rare sable antelope, the mighty eland, the stately koodoo, and the now extremely scarce roan antelope—the last-named extinct within the Cape Colony nearly these hundred years.

The name gemsbok is a good example of the extraordinary infelicity of the early Cape Dutch in naming their beasts of the chase. These old-world Batavians appear to have had about as much knowledge of natural history as of the electric telegraph. The largest antelope of the torrid regions of the Cape they called the eland or elk; the gnu they named the wildebeest (wild ox); the caama, the hartebeest (stag-ox), and so on down the whole gamut of wildly grotesque nomenclature. The oryx they christened gemsbok, which signifies, literally, a chamois, perhaps the most uncouth and far-fetched of the category. The gemsbok, whose Dutch name has clung to it these 200 years, is an antelope of striking characteristics. In size rather larger than an ass—the adult male standing some three feet ten inches at the shoulder—in general colour it is of a greyish buff. Its horns, straight, sharp-pointed, and over three feet long, are deadly weapons of defence; its form is robust and square, and its port very noble. But its most beautiful point is its head, which is of a pure white, painted with eccentric black markings that exactly and singularly resemble a complete headstall. The breast, stomach, and legs are pure white. A full and beautiful eye, a tuft of thick black hair upon the chest, broad black bands painted upon its back and sides, an erect mane reversed, and a long black switch tail, that sweeps the ground, all combine instantly to mark out this magnificent creature from its fellows, upon the broad plains whereon it has its habitat.

As a beast of the chase the gemsbok stands almost unrivalled. It is possessed of amazing swiftness and bottom, and when turned to bay, its sharp horns and determined charges render it an exceedingly awkward customer should the rifle not be held straight. Of all the antelopes, with perhaps the exception of the Zwart-wit-pens (the sable antelope), the oryx alone will face the lion;

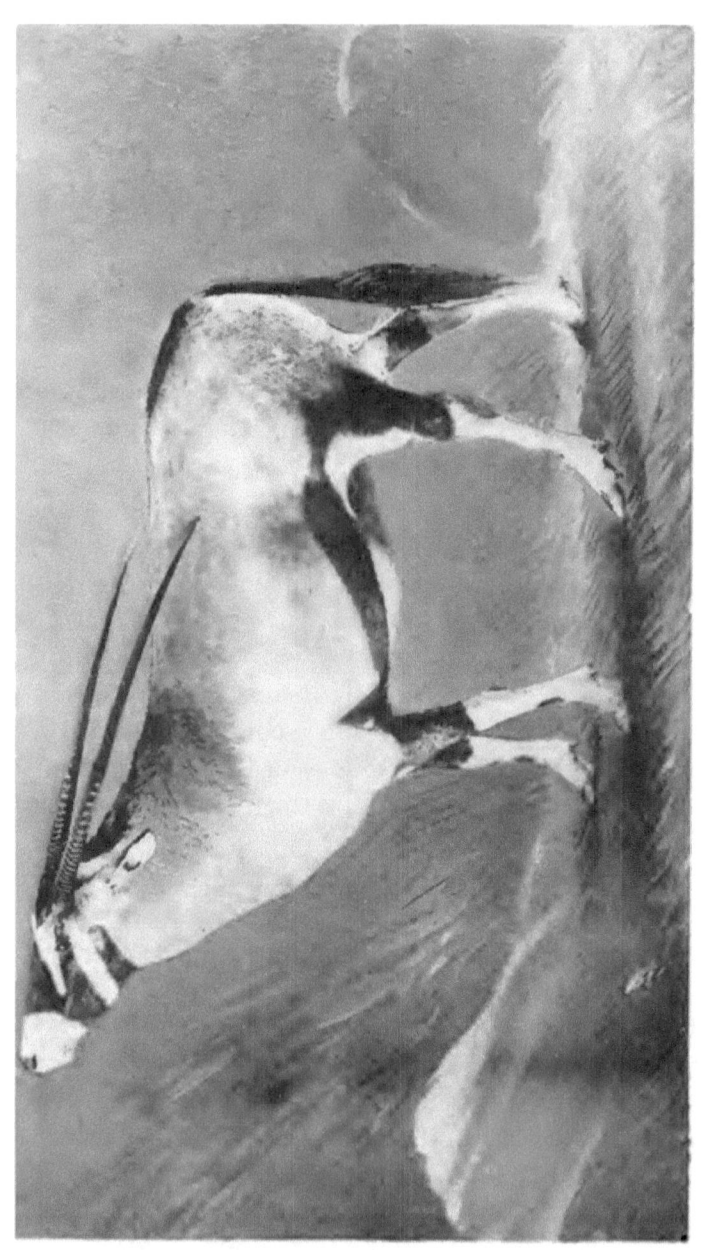

THE GEMSBOK OR ORYX

and it is an undoubted fact that not seldom have the carcasses of the lion and the gemsbok been found rotting on the plains together, the lion, impaled in its spring, firmly fixed upon the sharp horns of its quarry. An old-world Boer, who hunted in Cape Colony so far back as the end of the last century, when elephants, rhinoceroses, and other great game were abundant, described from personal observation to a friend of the writer, the singular method in which this antelope will often receive the attacks of lion and leopard. When danger threatens, it throws itself to the ground, and presents its tremendous horns to every point of attack, occasionally slightly shifting its position to enable it to do so. The eyes are prominent and set high in the head, and greatly aid to this system of defence. The leopard and lion, who never attack this antelope unless hard pressed by hunger, have been known to retire baffled from these trials of skill. Occasionally, no doubt, the great cats come off victorious, by stratagem or surprise.

Cornwallis Harris, Gordon Cumming, and other mighty hunters write in rapturous terms of the beauty, speed, and grace of the gemsbok. Andersson and Selous are, indeed, not so convinced of its superlative merits as regards speed, but it must be remembered that seasons, feeding, and locality have much to do with the condition of this, and, indeed, of all other antelopes. There are men, old interior hunters, who will tell you that to see the rare gemsbok ranging in its pride over its own primeval karroos, or the jet black sable antelope in its mountain fastnesses, is to them as keen a pleasure, even, as to gaze on the wondrous Victoria Falls of the Zambesi itself.

But, perhaps, the chief reason that renders the oryx of interest to Englishmen and Scots, lies in the fact that it is undoubtedly the original of the famous unicorn. Its colour, size, and shape, all point to this conclusion; and the horns, when seen in profile, appear as one. The ancients seem merely to have invented the forward set of a single horn, to have added a few other touches to the oryx, and to have thus furnished forth the unicorn to puzzle succeeding ages. Ctesias, physician to Artaxerxes Mnemon, who described it, *circa* 400 B.C., seems to have been the originator of this fable.

He describes it as a wild ass (*Onos agrios*), and from him doubtless has sprung the cloud of mystery in which the unicorn has been enveloped. Aristotle speaks of the oryx and the Indian ass as one-horned; Pliny follows Aristotle. Loborn, in his history of Abyssinia, describes the unicorn as resembling a beautiful horse. Other ancient authors of Greece and Rome refer to this mythical creature as of the size of a horse, and having the appearance of that animal, with one straight horn, of from one-and-a-half to two cubits in length, planted upon the forehead. Further interesting details had grown with the ages. The body was now described as white, the head red, the eyes blue; so swift was the unicorn that no horse could overtake it. A legend arose that the unicorn could only be captured by the aid of a virgin. By some ancient writers, the rhinoceros is clearly intended in the description of the unicorn. The English Bible in the Book of Job clearly follows this idea. In those wonderful descriptions of Job, the idea of enormous strength depicted in the words, "Canst thou bind the unicorn with his band in the furrow? or will he harrow the valleys after thee?" can only refer to the mighty bulk and gigantic power of the rhinoceros.

By some it has been contended that the unicorn of heraldry derives its origin from the Crusaders, who are said to have brought home accurate descriptions of this animal of mystery. But the Crusaders could only have obtained their information from those Eastern sources to which I have referred, or from Egyptian or Persian monuments. From monuments at Persepolis—the "Takti Jamshid" (throne of Jamshid), captured and destroyed by Alexander the Great—it would seem that the oryx, with a single horn set forward upon the forehead, is clearly intended in the Cartozonon of the ancient Persians. It is true that the wild ass, furnished with a long horn, is actually described; but the size, the colouring, the erect mane and switch tail, and the clean wiry limbs, are notably similar in the oryx and the wild ass.

In short, the investigations of distinguished naturalists and historians clearly establish the fact that from the monuments of old Persia and of Egypt springs the unicorn, and that this long-fabled creature is no other than the oryx with the horn set forward, and, no doubt from the profile view of the animal, represented as single.

Before the Union, two unicorns stood as supporters of the Scottish arms. The animal now forming the sinister supporter of the Royal Arms of Great Britain is "a unicorn argent armed, crinéd and unguled or, gorged with a coronet composed of crosses, patée, and fleurs-de-lys, with a chain affixed passing between the forelegs, and reflexed over the back of the last."

The picture of the gemsbok, facing page 386, is reproduced from a photograph of a specimen at the Natural History Museum. This, perhaps the finest example ever brought to England, was shot by Mr. F. C. Selous, on the Botletlie River, North Bechuanaland. I am indebted to Dr. Gunther for special permission to photograph this matchless specimen.

When in Cape Colony, I was greatly interested to hear that the singular and beautiful gemsbok, the Cape oryx, yet lingered in the wild desert country of North Calvinia, where its shyness, swiftness, and the fact of its existing in a country almost destitute of water, yet gave it sanctuary from the too relentless hunter. The Game Protection Act, passed a year or so since, gives the gemsbok and the hartebeest complete immunity from the gunner for the space of three years, during which time it is devoutly to be hoped that both these fine antelopes may increase and multiply, long to adorn the parched plains of the Old Colony. It may be noted that the oryx in slightly varying forms occurs in North-Eastern and North-Western Africa, and in Arabia. The gemsbok is now difficult of access, but if the sportsman seeks a happy hunting ground and worthy quarry, where yet this stately creature freely roams, he may reach Kimberley by rail in thirty-six hours from Cape Town, and thence the Kalahari desert, its mysteries and its treasures, lie not very far distant.

Chapter XXII.

THE EXTINCTION OF THE TRUE QUAGGA (EQUUS QUAGGA).

IT is a melancholy thing to chronicle the final disappearance from its ancient primeval habitat of one of the most beautiful forms of animal life; but there is now, I fear, no longer any reasonable doubt that the true quagga—quacha of the Hottentots—*Equus quagga* of Linnæus—must be numbered in the increasing catalogue of extinct creatures.

During my former stay in Cape Colony, I made many inquiries in various parts of the country as to the date of the final disappearance of this animal from within the colonial limits, and from the information I then obtained, I should be inclined to place its extinction south of the Orange River, between the years 1860 and 1865. Cape Colony is a wide territory, and its remote parts are to this day almost as unnoticed and unknown as they were at the beginning of the century; and for this reason, it is very difficult to place exactly—even to a few years—the actual extinction of its disappearing fauna. In the Orange Free State, where it roamed, not so very many years since, even more freely than of old in the Cape Colony, the quagga lingered to a much later period; but even there it would seem for some years past to have become quite extinct.

In this opinion I am fortified by Mr. F. C. Selous, who told me only a few months since, just before his last return to Africa, that he had not heard of a true quagga in the Free State for some years, and doubted not that it had become completely exterminated. Other South African informants confirm this opinion.

It may be well, before proceeding further, to remind readers that among all South African interior hunters and traders, Burchell's zebra—called by the Boers, from its fuller markings, the bonte quagga (literally the pied or spotted quagga), in contradistinction to the now extinct true quagga—is also invariably, if erroneously, called a quagga. In this way a good deal of confusion has arisen, and many people, no doubt, still suppose that the rarer true quagga is yet plentiful in the distant territories beyond the Orange River.

I find that among hide-brokers and others in the skin and leather trades, this confusion has also arisen, and that the skins of the Burchell's zebra, which still arrive in this country in large numbers from East Africa, are invariably classed as "quaggas."

The true quagga, as a matter of fact, was easily to be distinguished from Burchell's zebra, by the paucity of its stripings, which extended only to the centre of the barrel, and by its general body colour —a rich rufous brown, fading to white underneath. Burchell's zebra, on the contrary, is completely striped all over the body (save as to the legs), while its body colour is yellowish sienna. The true zebra (*Equus zebra*), formerly the rarest of the hippotigrine group, differs widely, again, from both the above-mentioned animals. In this species, as I have before pointed out, the body colour is silvery white, the stripings are black, and extend all over the body and legs, while the ears and tail are far more asinine. It may further be noted that, amongst the zebras called Burchell's, a variety exists—known as Chapman's variety—in which the legs are striped as completely as in the true or mountain zebra (*Equus zebra*). This variety is comparatively scarce, but not so scarce, Mr. Selous tells me, as is generally imagined. In equatorial Africa, indeed, it would seem to be fairly common.

It is, I think, unquestionable that, so far as mere beauty is concerned, Burchell's zebra (the zebra of the plains), the commonest of the group, carries away the palm. The type is more equine, the ears are smaller, and the tail more bushy than in the true zebra; while the colouring is, to the average eye, richer and more attractive. Thus much I am free to admit. On the other hand, I am bound to say of the true zebra that, viewed as I have viewed it in Cape Colony, revelling in the wildest, most rugged, and remotest portions of its own-loved mountain habitat, there are few prouder and more beautiful objects in animate nature.

Burchell's zebra, of which there are always some good specimens to be seen at the gardens of the Zoological Society, though for the last hundred years not found south of the Orange River, has a widely extended range over all Africa, from the Orange River to far north of the equator. The true zebra would seem to be confined mainly to the extreme north and south of the continent; its peculiar and most favoured habitats being the Cape Colony, and

Namaqualand and Damaraland in the south, and Abyssinia and Shoa in the north.

The range of the true quagga, the subject of this chapter, was even more arbitrarily defined. This animal, formerly so abundant upon the far-spreading karroos of Cape Colony and the plains of the Orange Free State, appears never to have been met with north of the Vaal River. Its actual habitat may be precisely defined as within Cape Colony, the Orange Free State, and part of Griqualand West. I do not find that it ever extended to Namaqualand and the Kalahari Desert to the west, or beyond the Kei River, the ancient eastern limit of Cape Colony, to the east. In many countries, and in Southern Africa in particular, nothing is more singular than the freaks of the geographical distribution of animals. A river, or a desert, or a belt of sand or timber—none of which, of themselves, could naturally oppose a complete obstacle to the animal's range—is yet found limiting, thus arbitrarily, the habitat of a species. I have thought a good deal over this circumstance, and I confess to being completely puzzled for a satisfactory explanation of it, more especially in the cases of the zebra, Burchell's zebra, quagga, giraffe (said, in common with Burchell's zebra, to have been never found south of the Orange River), the blessbok, bontebok, black wildebeest (white-tailed gnu), and other antelopes.

As the quagga may now be fairly considered as absolutely extinct as the dodo—for alas! although easy enough to destroy, no human skill can ever restore it to its natural haunts—it may not be out of place here to record its exact description, taken from Cornwallis Harris's " Portraits of the Game and Wild Animals of Southern Africa." "The adult male stands four feet six inches high at the withers, and measures eight feet six inches in extreme length. Form compact. Barrel round. Limbs robust, clean and sinewy. Head light and bony, of a bay colour, covered on the forehead and temples with longitudinal, and on the cheeks with narrow transversal stripes, forming linear triangular figures between the eyes and mouth. Muzzle black. Ears and tail strictly equine; the latter white and flowing below the hocks. Crest very high, arched, and surmounted by a full standing mane, which appears as though it had been hogged, and is banded alternately brown and white. Colour of the neck and upper parts of the body dark rufous brown, becoming gradually more fulvous, and fading off to white behind and underneath. The upper portions banded and brindled with dark brown stripes, stronger, broader, and more regular on the neck, but gradually waxing fainter, until lost behind the shoulders in spots and blotches. Dorsal line dark and broad, widening over the crupper. Legs white, with bare spots inside above the knees. Female precisely similar."

Sir W. Jardine describes the quagga as equal or superior in size to the Burchell's zebra, and as "still more robust in structure, with more girth, wider across the hips, more like a true horse; the hoofs considerably broader than in the true zebra, and the neck full; the ears rather small." The same writer speaks of it as best calculated for domestication of the zebra group, both as regards strength and docility. He instances the facts, that the late Mr. Sheriff Parkins used to drive a pair of quaggas in his phaeton about London, and that he himself had been drawn by one in a gig, "the animal showing as much temper and delicacy of mouth as any domestic horse."

Barrow, in his excellent "Travels into the Interior of Southern Africa" (1797-8), writes of the quagga as "well-shaped, strong-limbed, not in the least vicious; but, on the contrary, soon rendered by domestication mild and tractable; yet, abundant as they are in the country, few have given themselves the trouble of turning them to any kind of use. They are infinitely more beautiful than, and fully as strong as, the mule, as easily supported on almost any kind of food, and are never out of flesh."

On the other hand, although undoubtedly, if taken young, more easily tamed than the fierce true zebra, in its wild state the quagga must have been an extremely awkward customer when wounded. I have heard numerous anecdotes from old colonists of the ferocity of this animal and the zebra in the wild state. Cornwallis Harris instances the death of a native servant, whose skull was completely smashed by the kick of a quagga, and mentions a narrow escape of his own; and he further speaks of his having seen "a wretched savage, every finger of whose dexter hand had been stripped off by the long yellow teeth of a wounded male."

The courage of the quagga, reputed by Jardine "to be the boldest of all equine animals, attacking hyæna and wild dog without hesitation," was, in the old days, to some extent taken advantage of by the Dutch colonists, who are reported to have used them in a domesticated state " for the purpose of protecting their horses at night, while both are turned out to grass." I have not been fully able to satisfy myself that this account of Jardine's was correct. The Boers' ancient habit of kraaling all their flocks and herds, to insure them against the nocturnal attacks of dangerous animals, would, I think, militate against the adoption of this plan to any great extent. Still, in some instances, it may have occurred.

The vast karroos of the Cape Colony in the earlier half of this century, and the plains of the Orange Free State until about the year 1860, furnished the most extraordinary spectacles of teeming animal life that the world has ever seen. At one time the supply of game, which literally swarmed upon these plains, jostling one another in actual millions, seemed practically inexhaustible. No other part of the world could ever vie with these favoured regions in the wealth and variety of their fauna. Elevated table-lands, a singularly healthful climate, and abundance of pasture, seem to have here united to attract and sustain an unparalleled collection of animal life. Brindled and white-tailed gnu, Burchell's zebras, quaggas, springbok and blessbok, and the ever-attendant lion, formed the main body of all these beasts of the chase; ostriches, elands, hartebeests, gemsbok, and bontebok, furnished ample battalions in support. Old colonists have often spoken to me of the quantities of game thus existing as being positively inconceivable; but improved arms of precision, and the ruthless skin-hunters, have done their work, and well nigh all these crowds of game have vanished. In the Cape Colony, the beautiful springbok yet remains in considerable abundance to remind one of the past; but the Free

State is now almost entirely denuded of its former glorious fauna. Nature, upon these crowded plains, seems to have provided periodical and ravaging diseases to keep down the superabundance of life, and the *brand-sickte*, or burning sickness, amongst others, killed off thousands of game at various seasons. It is a singular circumstance that, although inhabiting the same plains in the Orange Free State, the quagga was never to be found mixing with its near relative, the Burchell's zebra. It was, however, curiously enough, almost invariably found associating with the black wildebeest (the white-tailed gnu, Connochætes gnu, now also becoming extremely scarce) and the ostrich. The Burchell's zebra for its part is, and was, usually seen in company with the blue wildebeest (brindled gnu, Connochætes gorgon); and the ostrich would seem to have been a companion common to both animals.

The downfall of the quagga first began, as it ended, at the hands of the Boers. In the old days, the Cape Dutch farmers, to save their flocks and herds, which they considered far too valuable to waste on the feeding of their slaves and retainers, usually shot the quagga for this purpose. The oily yellow flesh of this animal, disdained by themselves amid the abundance of more toothsome game, was set apart for the food of their Bushmen and Hottentot servants, and Mozambique slaves, and very large quantities were thus annually sacrificed.

Although this practice gradually thinned it off the Great Karroo, even so lately as Gordon Cumming's day (1843) the quagga was found upon the northern plains of the Cape Colony in considerable numbers. Upon the Great Karroo itself, the last quaggas ever seen were shot near the Tiger Berg, a solitary mountain, springing from the plains not far from Aberdeen, in the year 1858. There were two then remaining, and my informant, with whom I stayed while in Cape Colony, near this very spot, well remembered their being shot. Further north in the Colony, as I have said, they lingered to a still later date. Their skins were used for grain sacks, and I have seen old quagga-skin sacks still in use in a Dutch farmhouse.

Some twenty or twenty-five years back the Boers of the Orange Free State suddenly awoke to the fact that in the skins of the quagga and Burchell's zebra and other game they possessed a mine of wealth, and they at once set about the task of exterminating these animals. The mine took some years to work out; but the task is now complete, and many a farmer looks back with keen but unavailing regret to the glorious hunting veldt of the past.

In the hide trade the skins of these animals have been in great demand for years past, and the avaricious Boers were not slow to avail themselves of this demand. The bone-strewn plains of the Free State bear miserable testimony to the feverish haste and deadly methods of the skin-hunters of the past twenty years. The work was done with business-like skill and parsimony. Bullets were carefully cut out of the carcass for future use, and a sufficient number of skins having been prepared, the waggons were loaded up and slowly driven down country to the coast.

From the point of view of the naturalist, the final disappearance of the true quagga is a fact full of melancholy interest. That an animal so beautiful, so capable of domestication and of use, and to be found not long since in so great abundance, should have been allowed to be swept from the face of the earth, is surely a disgrace to our latter-day civilisation. No human effort can now recall this magnificent form—it is gone for ever, after an existence of untold thousands of years upon its spacious plains. One can only hope that the African elephant, the eland, and other useful creatures now fast disappearing, will not be suffered to vanish thus miserably.

I find that the true quagga has not often figured in the collection of the Zoological Society. There appear to have been but two specimens shown there: one, a female, purchased in 1851; the other, a male, presented by Sir George Grey (then Governor of the Cape) in 1858. These specimens have, of course, long since disappeared.

I do not remember to have seen a stuffed specimen of the quagga in the Natural History Museum at South Kensington, although I believe a skin or skins formerly existed in the British Museum.

The name quagga is obviously a corruption of the Hottentot name for this animal, "quacha," derived from the animal's call or neigh. Pringle, the bard of South Africa, recalls this peculiar neigh in his poem "Afar in the Desert":

> And the timorous quagga's wild whistling neigh
> Is heard at the fountain at break of day.

Old hunters and colonists have described to me the curious single-file march of this animal when moving about its native plains, and the squadron-like wheel of the troop when disturbed. Cornwallis Harris seems to have noticed this *trait*. He writes thus picturesquely of it: "Moving slowly across the profile of the ocean-like horizon, uttering a shrill barking neigh, of which its name forms a correct imitation, long files of quaggas continually remind the early traveller of a rival caravan on its march."

Never more, alas! will this fleet and handsome creature scour its primeval karroos, or proudly pace in regular formation, or with its fellows wheel and charge like a regiment of cavalry. It has been the first of the matchless fauna of South Africa to disappear. That its extinction may serve as some sort of warning to wanton and ruthless destroyers of game, whether Boer or British, or of any other nationality, is devoutly to be hoped.

Chapter XXIII.

THE FUTURE OF CAPE COLONY.

EVER since 1806, when a British force under Sir David Baird effected its capture, the Cape Colony has remained without interruption in the hands of England. Before that time it had been in our possession from 1795 to 1803, but had been restored to the Dutch at the Peace of Amiens. But although, in point of years, the Cape occupies an ancient position among our colonies and dependencies, it has, for a variety of reasons, lingered tardily behind in the race of material progress and prosperity. India, Australia, New Zealand and Canada have, within these last forty years, advanced with amazing strides; but South Africa, with advantages in many respects equal, in some superior, to her sisters, has hitherto not taken that part in the world that she had the right to command. Camoens, in his "Lusiad," makes mention of Adamastor, the spirit of the stormy Cape, and it would almost seem that the grim phantom he portrays had really cast an evil spell over this fair country; for the Cape has been to the British during long years a distressful and disturbed possession. But all, or nearly all, the influences that have created these troubles have passed, or are passing away. There are unmistakable signs that the dawning of a new era of peace, enlightenment and progress, has come for South

VIEW FROM BACK OF TABLE MOUNTAIN

Africa, and it may be hoped that Adamastor and his malignant destinies have disappeared finally and for ever. It is not to be imagined that the Batavian Government had, before the coming of the British, an easy or unquestioned sway over the Boers of the Colony. On the contrary, the absurd and arbitrary restrictions of succeeding Governors had more than once goaded the rugged farmers of the distant frontiers into insurrection, or something very near it. In 1795, the colonists of Graaff Reinet and Swellendam actually expelled the landdrosts and declared themselves independent. It has been suggested that these farmers may have been attracted and imbued by the spread of revolutionary principles, at that time flooding France and other parts of Europe. To those who know the Boers and their isolation, this assumption is in the last degree improbable; these graziers of the wilderness were at that time, as indeed they are to this day, completely shut off from the outer world, and having no external communication with it whatever.

At this juncture the British Government came to the assistance of the Dutch authorities, and with the consent of the Prince of Orange, their Stadtholder, then a fugitive in England from the armies of Napoleon, took possession of the Cape in the name of George III.

Upon their entry into the Cape territories, the British found their new subjects surrounded on every side by a network of absurd and antiquated and most vexatious laws. There is, perhaps, no more striking example of the rooted antipathy of the Dutch colonists to any sort of change, however necessary or favourable to themselves, than the state of their marriage law at the end of the last century. At that time the settlers, even of the remotest parts of the Colony, were compelled to present themselves at Cape Town, in order to be examined whether there were any hindrance of relationship, blood-kindred, or the like, and to pass certain other forms. Whether considered from the point of view of decency, of expense, or of convenience, no more monstrous or barbarous law could have been devised. Two young settlers of the frontier districts, such as Graaff Reinet, wishing to be united in matrimony, had then to undertake together a waggon journey of five or six hundred miles, occupying a space of two or three months, over parched and devious deserts, and encompassed by dangers from wild men and wild animals. And this notwithstanding that pastors and landdrosts (magistrates), who might readily have performed the office of marriage, were to be found within their own provinces.

For further observations upon this oppressive and degrading law, and its frequent consequences for the woman, the reader may be referred to p. 152 of Barrow's " Travels into the Interior of Africa," published in 1801. It is hardly necessary to mention that this enactment was speedily repealed by the British upon their entry into the Cape. But this was only one out of a multitude of antiquated and absurd restrictions that still obtained in the Dutch South African possessions at the close of the Batavian rule. Thus, in 1780, at the very moment that a British fleet menaced the Cape, the following grant of citizenship was made to one Gous, a tailor, who had formerly been a soldier :—" He is graciously allowed to practise his craft as a tailor, but shall not be allowed to abandon the same, or adopt any other mode of living; but, when it may be deemed necessary, to go back into his old capacity and pay, and to be transported hence if thought fit."

Prisoners awaiting their trial were then confined in dark and noisome dungeons, their legs secured in stocks or by iron chains, for months before undergoing trial, whether innocent or guilty. "This," we are told, "often had an excellent effect; they became, through it, so docile and so mild, that they confessed all that was in their hearts." The condemned were sometimes broken on the wheel, or stripped naked and bound to a wooden cross, whereon first the right hand was struck off, then the head, after which the body was disembowelled and quartered. At the time the British Government abolished these barbarities, the Dutch Colonial Court of Justice urged their continuance "as proper engines of terror." It is small wonder then that the Batavian Government at the Cape—a weak and yet a brutally severe one—had succeeded, at the time of British interference, in goading the colonists into insurrection.

It is a curious circumstance that the Dutch rulers at the end of the last century were as alarmed at the prospect of immigration to the Colony as the Afrikander Bund party is at the present day. Quite at the close of the Dutch dominion, Governor Janssens, replying to a memorial, said: "With regard to your inclination to strengthen the Cape with a new settlement, we must to our sorrow, but in all sincerity, declare that we cannot perceive any means whereby more people could find a subsistence

here, whether by farming or otherwise . . . . .
When we contemplate the number of children growing up, we frequently ask ourselves, not only how they could find other means of subsistence, but also what it is to end in at last, and what they can lay hands on to procure their bread."

Such was the freezing polity, such were the numbing fetters by which the Batavian authorities sought to restrain the vigorous existence of the infant Colony!

It may be gathered from the few facts I have quoted that the Boers, although they have been frequent and persistent grumblers against British dominion and influence, had in reality little to thank their own Government for; and the very fact of their insurrection indicates that they would not long have peaceably endured such a state of affairs as existed in 1795. Worried and harassed as they had been by the petty and exasperating legislation of their Governors, it is astonishing indeed that they had tamely endured it so long. But the evils that men (and Governments) do live after them. Some of the earliest enactments of the old Dutch Legislature will indicate the spirit in which they afterwards governed, and the principles they instilled into the early settlers—principles that even now obtain far too much amongst the Boers. Thus lands were granted on yearly leases at the small fixed rent-charge of twenty-four rix-dollars (thirty-six shillings sterling) in any part of the country. It was also enacted that the nearest distance from house to house was to be three miles, so that each farm consisted of nearly 6,000 acres. Concerning this policy, Barrow, then secretary to Earl Macartney, the first English Governor of the Cape, afterwards widely known as Sir John Barrow, secretary to the Admiralty, observes, "The Government foresaw that a spirit of industry, if encouraged, in a mild and temperate climate and on a fertile soil, might one day produce a society impatient of the shackles imposed on it by the parent State. It knew that to supply its subjects the wants of life without the toil of labour or the anxiety of care; to keep them in ignorance and to prevent a ready intercourse with each other, were the most likely means to counteract such a spirit." This is a heavy impeachment, but it would seem to contain the elements of truth.

One of the main difficulties in developing the Cape Colony has been the extraordinarily tenacious hold retained by the Dutch population upon territory which they themselves have never attempted to utilize, preferring rather to exist in a state of primitive patriarchal life, and using, or rather misusing, vast tracts of rich land, which with irrigation and culture might sustain thousands of souls, for the depasture of their flocks and herds. This grip upon the land is, however, at length relaxing. A succession of bad seasons, added to an effete system of farming; the temporary ruin of pastures by years of overstocking, and other reasons, have compelled large numbers of the Dutch population to mortgage their ancient and unwieldy estates. These mortgages have not been met; the estates have been foreclosed upon, realised and thrown into the market, and fresh owners with newer and enlarged ideas have taken, and are taking, the place of the more ignorant and unenlightened of the Dutch farmers. In some instances the Boers themselves have happily awakened to the fact that they must either march with the times or be left stranded, and they too are farming more in accordance with modern practice.

During the first four years of the British occupation, old taxes were diminished and modified, and no new ones were imposed. Every production of the Colony increased greatly in demand and value. More than 200,000 rix-dollars of arrears of rent were remitted, as well as 180,000 rix-dollars of other debts. The laws and religion of the inhabitants remained undisturbed; a much greater share of personal liberty was enjoyed; property increased to double its former value; the paper currency, which had been devised by the Dutch Government as a temporary stop-gap and had depreciated 40 per cent., rose to par with specie; and £2,000,000 of money was brought over from England and set in circulation. The rent of houses became doubled, and the country farmer found his sheep worth twice their former value. "Four years of increasing prosperity, of uninterrupted peace and domestic tranquillity," says Barrow, "have been the happy lot of the inhabitants of the Cape of Good Hope." There is a striking analogy between this rapid advance under British influence at the Cape nearly ninety years ago, and the brief period of British rule in the Transvaal. In the latter country between 1877, when the Union Jack first floated there, and 1881, when it was, to the everlasting disgrace of Mr. Gladstone and his Government, hauled down, property trebled in value, markets were good, capital was abundant, and everything pointed to a rapid and enormous expansion of trade and industry. The Transvaal, however, threatens to become Anglicized against its will. Despite the endeavours of President Krüger and the old Boer party, thanks to the discovery of the gold-fields, a British population is steadily flooding the territory of the South African Republic, railways are being pushed steadily towards its borders, and a few years will in all human probability see that rich country far more pervaded by the Anglo-Saxon element than is the Cape Colony now, even if it be destined never again to add lustre to the British crown.

In 1806, when the Cape for the second time and finally* fell into our hands, its borders were much more circumscribed than they now are; the Great Fish River on the east, and an irregular line drawn from Buffalo River on the west to the site of the present town of Colesberg, forming its limits in those directions, while the Atlantic and Indian Oceans bridled it to the west and south.

At that period the total population was 73,663, of whom some 27,000 at most were of European descent. Since that time the area of the Colony has been enormously increased. Bushmanland, Little Namaqualand, British Kaffraria, Griqualand West, and finally Kaffraria itself, have one after the other been added, forming a present total area of 213,917 square miles—nearly double the extent of Great Britain and Ireland—and having a total population of 1,377,213 souls, of whom considerably more than 340,000 are Europeans or of European descent. British and Boers are at present fairly evenly divided in the Colony, the latter probably exceeding the former by some 30,000.

The principal towns of the Cape Colony and their populations are now about as follows:—

Cape Town	50,000 inhabitants.
Kimberley	28,000 ,,
Port Elizabeth	16,000 ,,
Graham's Town	10,000 ,,
Paarl	6,000 ,,
King William's Town	5,000 ,,
Graaff Reinet	5,000 ,,
Stellenbosch	5,000 ,,
East London	5,000 ,,

Of these, Kimberley has sprung into existence since 1870; Port Elizabeth, Graham's Town, King William's Town, and East London practically since 1820; Graaff Reinet, Paarl, and Stellenbosch have chiefly increased since that time; while Cape Town owes its considerable population to British influence and British trade. These figures are fairly encouraging, but, compared with the populations of Australian and New Zealand cities, they are as naught. The European population of the country districts shows to even less advantage than in the urban.

Under British control the Colony advanced steadily, if slowly, until 1855. From that time until 1876 progress was more rapid; the labours of the Albany settlement of 1820 have borne fruit, after years of hard and uphill struggling. From 1870 dates the diamond industry, while the ostrich mania was at its highest from 1875 to 1880. From 1879 to 1886 a period of profound stagnation intervened. Every one had gone mad over ostriches. Farmers and traders cast aside their ordinary occupations and embarked in "birds." Feathers rose to £50 and £70 per pound, and breeding birds to £200 and £300 per pair. At length, in 1881, the crash came, prices fell suddenly and rapidly, and ruin, depression, and stagnation ensued. But the value of ostriches has reached its lowest level, and the industry may now be carried on profitably at the low prices that obtain for stock. Meanwhile, in spite of depression, the quantity of feathers exported has steadily risen, in itself a healthy sign.

From 1865 dates the first production of mohair, from the importation of Angora goats; the value then exported being £368. In 1883 the value reached £271,804—a truly astonishing advance. Angora farming is now a settled and increasing industry of the Colony; the clip of hair from these beautiful animals comparing favourably with the choicest produce of Asia Minor, whence the original stock was imported.

Here are the values of a few of the chief exports of the Colony in recent years:—

	1865.	1870.	1875.	1880.	1885.	1886.	1887.	1888.
	£	£	£	£	£	£	£	£
Copper Ore............	118,297	146,368	248,537	306,790	395,675	559,328	577,053	856,803
Corn, etc.	30,246	33,241	5,490	5,042	4,975	7,960	18,256	19,599
Ostrich Feathers	65,736*	87,074*	304,933*	883,632	585,278	546,230	365,587	347,792
Angora Hair or Mohair	368	26,673	133,180	206,471	204,018	232,134	268,446	305,362
Hides, Ox & Cow	9,724	21,710	38,964	26,691	128,915	117,872	96,756	66,652
Ivory	10,056	13,746	60,402	16,982	3,629	2,150	4,774	2,089
Diamonds	Nil	153,460	1,548,634	3,367,897	2,489,659	3,504,756	4,242,470	4,027,379
Goat Skins.........	60,621	126,112	158,404	107,428	103,209	104,894	99,923	109,068
Sheep Skins	79,160	87,240	147,842	172,264	192,631	174,225	169,981	198,107
Wine (ordinary)	24,499	13,887	12,817	11,966	14,558	21,593	17,190	18,378
Wool:								
Fleece washed	⎫	513,117	702,354	395,966	164,977	171,788	167,584	185,093
Scoured	⎬ 1,680,826	934,726	1,834,114	1,517,630	976,209	933,680	1,007,507	1,058,074
Grease	⎭	221,675	319,431	515,775	284,922	744,649	499,840	938,343

* 1865, wild feathers. 1870, principally ditto. 1875 and after, feathers from tame birds.

In addition to these products, aloes, salted fish, dried fruits, horses, horns, brandy, Constantia wine, and other articles are exported. Of the above, it is probable that in the future, corn, which the colonists have for years not only almost ceased to export, but have actually imported to the extent of from a quarter to half a million sterling annually, will be produced and exported heavily.

The ivory trade has dwindled to a shadow, in consequence of the almost complete extirpation of elephants south of the Zambesi. When it is remembered that soon after the discovery of Lake N'Gami, in 1849, 900 elephants were slaughtered by ivory hunters, within one short year, in that region alone, this is scarcely to be wondered at; but with the future continuation of Cape railways, as may be expected, from Kimberley towards the Zambesi, and perhaps beyond Lake N'Gami, to the western regions, as suggested by Mr. H. M. Stanley, a great Central and Western African ivory trade will be tapped.

Horses are reared at the Cape, of excellent quality and in great abundance, and there is no reason why they should not be exported at a profit in years to come.

Wine production, which would probably better repay the introduction of capital and enterprise than almost any other Cape industry, is likely, with improved methods, to increase very largely; for the soil and climate are even more favourable to the growth of the vine than in Australia. Yet Australia has, thus far, quite outstripped the Cape in the quality and quantity of its wines. This matter has, happily, now been taken in hand by the Cape Government, and experts from Europe are employed in teaching proper and improved methods. The old Dutch system seems to be chiefly remarkable for its dirt and carelessness.

Dried fruits were formerly exported to a greater extent than at present. South Africa is peculiarly fitted for the production of almost all kinds of fruit, and there is no reason why fruits, both fresh and dried, should not be exported at reasonable profit. If South America and Australia can export oranges, the Cape should surely be able to do the same. Cape oranges, grapes, peaches, apricots, nectarines, figs, apples, pears, loquats, quinces, guavas, and numerous other fruits are of fine quality.

Tobacco is an article that, instead of, as at present, being largely imported for the better class of smokers, might be exported to an enormous extent. The colonial production is about 3,000,000 lbs. per annum, a great portion of which is, however, used for sheep dip. The better qualities of Cape tobacco are habitually smoked by the farmers, English and Dutch, and command a ready sale. The Cape Government has, quite recently, engaged the services of an expert to impart information to the tobacco farmers. Hitherto, the fault seems to have lain, as in viti-culture, not in the growth, which in this favoured climate is simple enough, but in the curing and preparing for consumption.

The diamond industry, in Griqualand West, now an integral part of Cape Colony, has long been established on a secure and steady basis, and the output appears likely to be maintained for many years to come, at from £2,000,000 to £4,000,000 per annum. The returns received for 1887 show that the value exported in that year amounted to nearly four millions and a quarter. Moreover, it is not unlikely that, in process of time, other diamond-bearing localities may be discovered. These precious stones have been found in the Orange River, at least as far down as Prieska, to the north-west of the Colony, and there are many more improbable things than that rich diamond areas exist far away from Kimberley. Who would have dared to say, twenty years ago, that upon those arid plains, whereon a few miserable Griquas then herded their scanty flocks, Kimberley, and its enormous potentiality of wealth, would now stand?

In minerals the Cape is extremely rich, yet hitherto her territories have lain unnoticed and almost entirely unprospected. The copper mines of Little Namaqualand, south of the Orange River, are among the richest in the world, and are at present nothing like developed. Copper exists in many other localities in the north-west.

Coal has been found in many parts of the Colony, and no doubt will crop up in very many more. Notably in the Stormberg range excellent coal, which is now largely employed on the Cape railways, is obtained. Gold has, within the last two years, been discovered in considerable quantities at the Knysna in the extreme south, and a mining camp is now established there. Still more recently, within the last few months, other discoveries of this precious metal have been made in the north-west, near the Orange River, and in other places, and with the gold fever that has now laid such a hold upon South Africa, it is not unlikely that large and important auriferous areas may be found in other parts of the Colony.

Other minerals are plentiful, but as yet lie

unsought and undeveloped. Manganese, iron, lead, and other ores are abundant in many localities, and it is not improbable that the Orange River region—hitherto unknown and almost unexplored—will yield vast treasures of mineral wealth. In that magnificent, mysterious river are to be found diamonds, amethysts, cairngorms, garnets, quartz crystal, rose quartz, jaspers of many varieties, agates, chalcedony, and a host of other stones. Marble of fine quality and colour, crocidolite, asbestos, granite, freestone, sandstone, prehnite—a very beautiful green ornamental stone—and porcelain clays of the finest qualities are also freely distributed throughout the Colony, but have hitherto been almost entirely neglected. Salt pans have long been a source of wealth in various parts.

In truth it may be said that even to its own inhabitants the Cape—save in its pastoral sense—is to this hour a *terra incognita*, and that riches, in almost extravagant profusion, lie scattered about the land, waiting only for the coming of the capitalist and the prospector.

And with the development of these and other products, most assuredly to take place in the not very far distant future, an education, silent but wondrously effective, will be going forward amongst the native populations. Glance at the mines of Kimberley and Namaqualand. Think of the thousands of natives annually passing through an apprenticeship in the ways of regular labour at these semi-savage emporiums. To these centres flock representatives of the dark races from the uttermost parts of South and South Central Africa. From the far Zambesi, and even from beyond its broad waters; from the lands of the Damaras and Ovampos; from Matabele and Mashonalands; from the countries of the Amatongas, the Amaswazis, and the Zulus; from Kaffraria and Basutoland they throng. And not only do these races learn lessons of obedience, of the value of money, of self-control, and of many of the arts of peace; but above all they acquire a firm and lasting belief in the good faith of the British paymaster. Native hunters far up country will take unhesitatingly, and without misgiving, the letter of a British trader or sportsman for the payment of a sum of money, at a place hundreds of miles distant. For the development of progress and of peace among the barbarous tribes of Austral-Africa, as for the development of the immense natural wealth of the country itself, immigration, new blood, and the introduction of capital, properly directed, are alone required.

While a vast expansion of territory has been going forward at the Cape, there has not, however, compared with more fashionable colonies, been a corresponding influx of British blood and British energy. Australia, New Zealand, and Canada have been building and peopling great cities, rapidly filling up their back countries, and advancing, until quite recently, by leaps and bounds, in material wealth, while South Africa has plodded with painful and halting footsteps upon her devious route. What are the reasons for this disparity? Neither the climate, nor the quality of the land itself, can be adduced for one moment as sufficient explanation. The climate of Cape Colony is indeed one of the most magnificent and healthful in the world. There you may see in the descendants of the stout and stubborn breeds of Dutch and British examples of manhood unsurpassed in any land.

In that clear and bracing atmosphere the European constitution is invigorated instead of becoming enervated as in India, the West Indies, and other of our possessions. South Africa, in truth, is no drain upon the parent European stock, but rather reproduces, often on an increased scale, the bones and sinews of our northern lands. As for the quality of the soil, it cannot be urged that South Africa is poorer than her neighbours. Within the limits of the Cape Colony there are millions of acres of rich land now lying waste simply for lack of irrigation, of capital, and of emigrants. Hereafter I will touch upon the capabilities of this land in the way of grain producing, when cultivation and irrigation are brought into play. The climate, the flora, and in many instances even the terrains of Australia and Cape Colony, are strangely similar, yet has Australia far outrun her sister of the Cape in the race for wealth. First, I think, among the reasons for this want of progress may be placed the dragging influence of the Dutch population. As I have previously pointed out, the Boers were, and still are the principal landowners. But in recent years hard times have told sadly against them. There may now be seen in each week's issue of the *Cape Times* and other papers, frequent instances of the bankruptcy of Dutch farmers, necessitating usually a change of ownership in the soil. A silent if distressing revolution is thus being effected, probably for the ulterior benefit of the whole community. The Boers, except in a few favoured districts in the vicinity of Cape Town, inhabited by the older and

better class of Dutch families, have never made any considerable attempt to practise agriculture on a large scale. Their wine and brandy are through careless manufacture not of a character to recommend them to Europeans; their wool is too frequently sent to market in a wretched and filthy condition; and their tobacco, which might with care and attention vie with American growths, is grown and cured in a far too rough-and-ready manner, and only obtains a sale at extremely moderate prices within the Colony.*
The Dutch Afrikanders have been chiefly responsible for the prolonged and determined opposition to the passing of a Scab Act, an Act which at length carried through, will, it is computed, benefit the flock-masters to the extent of half-a-million annually. The Cape Dutch, indeed, cling with an extraordinary infatuation to the habits and ways of their forefathers. To this hour they will tell you that what sufficed for their fathers will suffice for themselves, and a book of travels of a hundred years since almost exactly describes their present customs and manners.

Yet in some respects has the Boer of South Africa been, as I think, a too heartily abused individual. He has his good points, as those who have had intercourse with him must admit. No one who recalls the stubborn and self-denying independence of those emigrant farmers, who in 1836 forsook the land of their adoption (a land passionately beloved and remembered by them even to this day), and trekked with their waggons and their oxen, their families and worldly goods, into the unknown wilds of the Orange Free State, the Transvaal and Natal,

---

* Boer tobacco sells at from one penny to sixpence per pound.

VIEW OF DEVIL'S PEAK, NEAR CAPE TOWN.

or the stirring and bloody fights that subsequently took place with Moselikatse and his hordes beyond the Vaal, and with Dingaan and his Zulus in Natal, can deny this strange people the qualities of determination, self-reliance, and a deep religious sentiment. They have been, it must be remembered, the pioneers of South Africa, and within these two hundred years they have done yeoman's service in opening up a barbarous and difficult country.

Moreover, I question if they have been altogether fairly treated by the English settlers. Naturally desiring to be left alone, they have been too literally set apart from the world, too severely left to their own devices; and there has been often too much scornful contempt of them as a stupid, an uncouth, and an inferior race. A little more sympathy, a little more friendliness, would, I am convinced, frequently have worked wonders with these simple and isolated farmers. A gentleman of my acquaintance, upon whose farms I stayed in the Colony, accomplished with the Boers of his neighbourhood, by the exercise of these qualities, what no Englishman had ever even attempted. He was the first British farmer, I believe, to induce the poorer of the Dutch Afrikanders to enter his service as overseers of his numerous outlying flock stations. But even he only won his way amongst them by dint of unwearied patience, and by entering their dwellings and conversing with them as if he took an interest in them. In time I believe the Boers will gradually adopt larger ideas, cease to dream only of hugging huge tracts of comparatively useless country, of flocks and herds, and of moody isolation, and will turn their attention, as they see Englishmen around them doing, to the fencing of their runs, to irrigation and the growth of grain, and to improved methods of treating their wool, wine, and tobacco.

The second great deterrent to Cape progress, and especially to its colonization and development, has been the constant succession of native wars that, until 1878, vexed the Eastern frontier. But it may be safely asserted that Kaffir wars are now things of the past. The Transkeian territories have been annexed to the Colony; the natives have settled down to the cultivation of their lands; and if they are properly looked after, and not suffered to be ruined in body and purse by the drink traffic, they will, while themselves greatly benefiting, indirectly prove a source of wealth to the Colony and the Mother Country. This deterrent, then, of Kaffir troubles may be considered as entirely removed, and it may be taken for granted that as vast numbers of natives have been for years past attracted to the diamond-fields by the demand for labour, as they have been thereby enriched and partly civilised by contract with Europeans, so the Kaffirs will, when labour markets are opened up by new industries, mining and otherwise, by irrigation and increased cultivation, tobacco manufacture and the like, cast off their ancient laziness, and enter in large numbers into the service of the colonists. Hitherto they have only been utilized as herds and grooms. In the Western province the Hottentots perform most of the work in the vineyards and corn-fields, but they are as a race, unhappily, almost completely ruined by the old custom, still obtaining among the Dutch farmers, of paying them partly in cheap and deleterious wines and spirits.

The Cape long suffered from the lack of railway communication, and, until the last few years, nearly every article of consumption and of commerce was dragged up and down country by the expensive and painfully laborious medium of the old-world ox-waggon. The first railway in the Colony was begun so recently as 1859. It was not until 1875 that any considerable progress was made in this direction. Within the last twelve years the advance in railway communication has, however, been extremely rapid. The lines from Cape Town and Port Elizabeth to Kimberley have been for some time completed, and the Colony has now some 1,600 miles of railroads, constructed at a cost of £13,407,385.

The Witwatersrand gold-fields of the Transvaal, destined, as Sir Donald Currie tells us, to become among the richest in the world, are already exercising beneficial influence on the Kimberley line, as witness the very marked increase in this last year's income of the Cape railways. It will not be long probably before further communication is opened up with the Transvaal, the Orange Free State, and the far interior (through Bechuanaland); and the Cape will proportionately benefit thereby. It is true that the Transvaal and its President have hitherto declined to entertain an extension into their territory from the south-west, and the Orange Free State Volksraad have postponed a proposed railway from Kimberley through their territory for the space of a year; but, after all, wiser councils may prevail, and it may yet be hoped that united action of the South African States and Colonies may be arranged. But whatever the event, if our colonial authorities are vigilant, will keep an eye on Swaziland and

Delagoa Bay, and wait for the rising tide and influence of English-speaking miners and men of business, now steadily flooding the Transvaal, the latter will ere long have their way in this matter. Last year (1888), an extension of the railway system from Kimberley to the Vaal River, and from Colesberg to the Orange River, was voted by the Cape Legislature, despite the opposition of Mr. Hofmeyr and the Afrikander Bund. Amatongaland, abutting on Delagoa Bay, has now been practically secured to England by treaty with the Queen Regent, and within these last twelve months we have wisely forestalled German and Transvaal encroachments in the direction of the Matabele and Mashona countries, north and north-east of the Transvaal, by the extension of our sphere of influence over those territories, and have thus secured to ourselves the future of the vast and fabulously rich regions lying between the Limpopo and Zambesi. In this matter our Colonial Office has moved in time, but there yet remains something which must be done quickly and done well. A hostile line and hostile tariffs from Delagoa Bay must at all cost be prevented, and a communication between the Cape railways and that port, *viâ* the Transvaal, will in due course follow. Sir Donald Currie, during his visit to the South African Republic in December, 1887, pointed out to President Kruger the immense advantages accruing to all the South African States and Colonies by such inter-communication, and it was believed that his words had weight with the bluff but acute Boer leader. But recent developments tend to show that Kruger has been playing a very astute game of his own. He has now manœuvred the Portuguese Government into such a position that the Delagoa Railway—for which he has been playing so long—was recently confiscated with the ulterior view of selling it to himself or his nominees. Meanwhile, aided by his Afrikander Bund friends of the Cape Colony, who temporarily shelved the Bechuanaland extension already voted, for a rival line through the Free State, he deluded the Cape colonists into the idea that he would assist in a line through the Free State to the gold-fields. Having gained the time he required for the development of his Delagoa Bay policy, the Cape party find themselves, after a year of waiting, left out in the cold, the Orange Free State Volksraad, acting with Kruger, having shelved the Kimberley-Bloemfontein extension for a year certain. It was a pretty scheme enough, but the firm attitude of the British Government in connection with the Delagoa Bay Railway Confiscation may probably render it after all futile.

The Cape Government, after being thus humbugged (there is no other word for it) by Kruger and the Orange Free State for a year, will now do well to turn their attention immediately to the Bechuanaland extension, their proper and legitimate outlet to the interior, for which an Act was passed a year ago. A British company is ready and willing to help them from the British Bechuanaland southern border northwards.*

Certain it is that the gold-fields of the Transvaal,

---

* Just as these pages go to press, a cablegram from the Cape announces that Sir Gordon Sprigg has arranged with the British South Africa Company for the immediate extension of the Bechuanaland Railway.

of Bechuanaland, of Swaziland, and of the still more distant and probably immeasurably richer country of Mashónaland, will, within the next ten years, entirely transform Southern Africa; as certainly the Cape Colony will, through her railroads and in other ways, undoubtedly share in the general prosperity now just dawning.

It is probable that in the near future, well-devised systems of irrigation, fencing, and tree planting, will have marvellously altered the face of the Colony. In the last few years these questions, which affect, not partially, but in an absolutely vital degree, the general welfare and the agricultural and pastoral progress of the community, have moved rapidly to the front. When the early settlers landed at the Cape, they found a land immeasurably richer in the vegetable productions of nature than at present. The mighty plains were covered with rich grasses and shrubs that have too often been overstocked and trampled to death by the farmers' flocks. Many of the mountains that now stand drear and barren were shaded by cedars, olives, and other trees, while every stream and water-course was margined by sturdy thorn trees, that had flourished and perpetuated themselves for untold centuries. The rainfall was then more abundant. But with a fatuous disregard for the future, the Boers set themselves to destroy hurriedly, and at haphazard, these bountiful gifts. Trees have everywhere been ruthlessly cut down, farms overstocked, the vegetation destroyed and trampled out, and grasses burnt off by periodical fires—a most short-sighted policy. The country has in consequence become year by year more desiccated, fountains and rivers have dried up or diminished, and the rainfall has been hindered or absolutely driven away.

As Mr. Gamble, late hydraulic engineer to the Cape Government, has remarked: "The wandering sheep-farmer goes on to Crown lands and cuts down the trees so that his goats may feed on the leaves; the Kaffir destroys thousands of saplings in the Kaffrarian forests to restore his easily and frequently burnt hut . . . . ; large areas along the Hart River and elsewhere have been denuded of the kameel dorn (acacia) tree to supply firewood for the machinery at the diamond-fields." . . . . . . "Erosions," caused by the incessant trampling backwards and forwards to kraal of useless and unwieldy flocks, "form into canyons thirty feet deep and upwards, and take off the surface-water and under-drain the land." Nature has in her turn retaliated, and the parched earth, deprived of the shade and moisture of its trees, and worn out by over-grazing, has proved comparatively valueless.

But happily new methods and new ideas are being brought into prominence by the better class of farmers, and these wasteful systems are gradually disappearing. The Irrigation Acts of 1877, 1879 and 1880 afford great facilities to the colonists, and although the Dutch farmers are from ignorance, inertness, and mortgage difficulties, slow to take advantage of them, they are gradually making way. Irrigation is, in truth, only just beginning to be properly appreciated at the Cape, yet already it has wrought wonderful results. A perusal of the Cape Irrigation Commission Report for 1883, and of the Cape Government Official Handbook for 1886, will demonstrate the extraordinary results attainable from even the simplest forms of irrigation: "In the district of Calvinia seventy bushels of wheat for one bushel sown can be relied on, while a return of one hundred and thirty bushels is not unfrequent" (Cape Handbook, p. 133). From my own knowledge of the grain farms on the Fish and Zak Rivers of Calvinia, I can assert that these figures are not exaggerated. It may be noted also that two crops yearly are obtainable (pp. 233, 234). From page 117 of the Irrigation Commission's Blue Book, I quote as follows: "At a farm called Nooitgedacht, on the Lower Oliphants River, the proprietor, thirty-five years ago, was unable to raise sufficient bread for his own consumption. At this moment it supports two hundred and sixty-five souls, whites, exclusive of the coloured population; and this is the result of irrigation carried on in its most primitive form, there being nothing beyond ordinary Boer dams made in the bed of the river, with furrows leading therefrom."

But instances, far more remarkable than this, of the extraordinary results obtainable from irrigation in this favoured climate, are scattered broadcast throughout the Irrigation Report. The principal attempt by the Cape Government at the storage of water on a large scale, has not long been completed, and the results, although at present necessarily imperfectly developed, already point to a gigantic success.

This work, the reservoir of Van Wyks Vlei in the Western province, was accomplished in a waterless tract of country—in Kayen Bult—a southern prolongation of the Kalahari desert. The catchment area is 460 square miles. The reservoir can hold 25,000,000,000 gallons, but it is never likely to be filled to overflowing. Last year no less than 4,000,000,000 gallons were contained. The Vlei had the appearance of an inland sea, and the Boers flocked from all parts of the surrounding country—in many cases from extraordinary distances—to witness the prodigy. This and other efforts of the Cape Government are likely to bring forth good fruit in coming years.

But in addition to irrigation by the catching and storage of water by reservoirs, other sources are available. Until quite recently, it was imagined that the Great Karroo could never be made available, save for the depasturage of the farmers' flocks. This mighty plain, waterless though it apparently is, has a marvellously rich soil, sun-baked, it is true, yet none the less fruitful where water can be brought to bear. It has long been known that streams of water, arrested by igneous dykes—called by the Boers yzer klip kopjes (iron stone ridges)—run plentifully beneath the surface of the plains. These are now being tapped and made use of. Windmills and wells are beginning to appear upon the karroo with highly successful results, and will undoubtedly now rapidly multiply. There are enthusiasts who predict that the whole of the Great Karroo plateau will one day smile with crops of waving corn; and though this may be too sanguine an estimate, yet many more improbable things have come to pass.

It can scarcely be credited that for years the Cape colonists have been importing wheat from abroad at the rate of £300,000 to £600,000 per annum, and this while thousands of acres of rich land within their borders have been lying wasted and unemployed. The thing is incredible, yet literally true. In the last year or two some diminution in this importation has occurred, but it is a standing disgrace to Cape agriculturists and Cape rulers that even one bushel of wheat should have to be imported into a country so naturally adapted for its production. It is not too much to hope that within a few years the Colony will not only cease to import, but actually be able to export at a profit. As to its quality, Cape wheat cannot be surpassed by any in the world; the better qualities, on the rare occasions upon which they have appeared, have commanded a ready sale in the London market for the finest kinds of pastry flours. Barley is prolific and good; oats and mealies (Indian corn) also furnish fine crops.

I have in my mind a well-known karroo farmstead, which supplies an excellent example of what may be accomplished on karroo land by the conservation of water and systematic irrigation. There a mass of foliage now decorates the once barren and forbidding earth. Umbrageous trees, flowering shrubs, roses and other beautiful plants, welcome the wearied traveller. Vineyards, garden-ground, corn and lucerne lands, make the earth rejoice. Hundreds of fruit trees flourish, some of them (fig trees) covering twenty-five by ten yards of ground. Mulberries rear themselves to heights of forty and fifty feet. Oranges, apples, pears, peaches, plums, nectarines, loquats, lemons, quinces, pomegranates, walnuts, almonds, cherries, and strawberries, all thrive amazingly.

To the credit of the colonists, there appears likely to be a rapid increase in the system of tree planting. Within the last few years, Arbor Day, a celebration borrowed from America, has been observed throughout the land, and now each year on this anniversary, which is observed as a general holiday, thousands of trees are planted by men, women and children. This is a wise and most praiseworthy custom, and succeeding generations of Afrikanders will doubtless bless the forethought of their ancestors.

Much, however, remains to be done in this way, and long years must elapse before the destruction of two centuries can be repaired and the forests of the Colony regenerated.

Another rapidly growing system which also promises to work wonders is that of the fencing of sheep and goat runs. In the old days, upon every pastoral farm, the flocks were kraaled at night in inclosures, as a protection against wild beasts, and in the morning led to their pastures in charge of Kaffir or Hottentot herds. The disadvantages of this system are sufficiently obvious. The sheep or goats have insufficient rest, and often improve but little in flesh or wool. The distance to pasture daily increases; great portions of the veldt are worn and destroyed by the incessant trampling to and fro to kraal; the kraals themselves become unspeakably filthy from the deposits of years, and harbour and propagate various diseases; while the manure, instead of being scattered about the veldt, thereby improving the pasturage, is wasted, or only partially utilized as fuel in its sun-dried form. This custom still obtains on very many pastoral farms, especially among the slow-moving Boers, its originators.

Picturesque it certainly is, as I can testify. On a hot African evening, as the sun goes down redly upon the karroo, it is a curious and most interesting spectacle to watch first the clouds of dust that herald the approach of the parched and weary flocks. Presently the troops of sheep and of snow-white silky-fleeced Angora goats emerge from these clouds, and, following their Kaffir herd-boys, trot to the dam to lave their burning throats. Then they turn away and file quietly into their respective kraals for the night. But the custom is rightly doomed; the evils of overstocking, the destruction of pasturage, the propagation of disease, and the decrease of flocks, at length became so serious as to call for a Government Commission. The fencing of runs was then suggested, and during the last ten years amazing progress has been made in this direction. The predatory wild beasts, formerly so destructive to the farmers' flocks, are rapidly disappearing, and the leopard (only found in mountainous regions), the jackal, occasional hyenas and wild dogs, and the wild cats alone remain. These survivors are being vigorously proceeded against by means of poison and trapping, and will eventually disappear, as have the lion and practically the hyena. Huge farms, from 20,000 to 30,000 acres (and even more) in extent, are now completely fenced, and the flocks graze by day and night in peace and comfort. The farmer can move them from one inclosure to another at his pleasure, the veldt is allowed rest, and a reserve is thus provided in time of drought. A compulsory Scab Act recently passed, after years of bigoted and factious opposition, provides another invaluable safeguard for the flock-masters.

But the Cape Colony, although emerging rapidly from its troubles and its strangely antique ways, is sadly lacking in a strong and wholesome tonic. Despite the failure of large numbers of the Boers, for some years past the old Dutch party has been rising in power (a power but too vigorously stimulated by the Transvaal War, the disaster of Majuba Hill, and Mr. Gladstone's inglorious surrender), until Cape ministers, though nominally Britons and representatives of British ideas, obey but the dictates and carry out the political traditions of the Afrikander Bund. It is difficult for us in this country to fully comprehend such a state of affairs. Under this malignant and overpowering influence, immigration has entirely ceased, and the Dutch party loudly declare that they will neither tolerate more settlers, nor the ultilization of the vast tracts of land now lying fallow and neglected. They tell us in fact that there is enough for themselves and no more. Mr. Arnold White, in his experiences in connection with his model settlement of Wolseley in British Kaffraria, a settlement boycotted and almost strangled in its infancy by officials of the Cape Government, can bear testimony to the extraordinary difficulties thrown in the way of organised attempts at colonization.

It is high time that Englishmen knew these things, and that they were ended once and for all. It cannot be tolerated that a rich field of emigration, such as the Cape, should lie longer fallow and unknown.

The tonic of new blood—British blood—and new ideas is urgently needed at this moment in Cape Colony, and the time is ripe for a fresh and considerable transference of colonists to our South African possessions. Whether such a transference shall take the form of a State-aided scheme, or of a scheme or schemes backed by private wealth, is a matter for mature consideration.

Since we took possession of the Cape territory, there has been only one considerable plantation of British blood upon its shores, and that a strikingly successful one. The Albany settlers of 1820 have proved themselves a tower of strength to the Colony. The five thousand emigrants sent out by the British Government, at a cost of £50,000, have completely transformed the Eastern province, now by far the richest and most advanced portion of the Colony. The plantation of a number of the German Legion in British Kaffraria, after the Crimean War, proved a similar success. But with these exceptions—exceptions of singular encouragement —no systematic emigration to the Cape Colony has taken place from this country. It is time that the tide turned, and this fair inheritance of ours became utilized.

As a health resort—boasting a climate unsurpassed —as a field for the farmer, whether agricultural or pastoral, the Cape needs only to be more known to be better appreciated. At this day a farmer in England, whether his capital be a few hundreds or a few thousands, now finds it well-nigh impossible to make both ends meet. A farmer with a capital of a few hundreds can live at the Cape in comfort and independence, under skies fairer a thousand times, and in an atmosphere a thousand times clearer and more exhilarating than at home. He may surround himself quickly with almost every fruit that man can desire, and will find his flocks and herds speedily increase.

He may buy good land for from five to ten shillings per acre, not an extravagant price it must be admitted. Game of all kinds, from the magnificent bustards and francolins to the springbok—gracefullest of the antelopes—still remains in abundance, so that sport, dear to every true British heart, is not lacking.

There are, as I have endeavoured to show, indications that the good time of this neglected possession is at hand. I believe that this generation will see the Cape Colony, with its water supplies—hitherto so wasted—enlarged and secured; its terrain re-afforested; its rich waste lands and almost unknown back-country laughing under lusty crops of corn, of bountiful vineyards and of fruit; its boundless pastures fenced and nurtured, and rendered a hundred-fold more profitable; its industries regenerated and enormously recruited. I see planted there a fresh and ardent Anglo-Saxon population, happy, prosperous and contented, mingling with the descendants of the ancient Dutch settlers on terms of friendship and good will. These things are not as the dreams of a dreamer, but can and will be vigorous and enduring realities.

www.ingramcontent.com/pod-product-compliance
Lightning Source LLC
Chambersburg PA
CBHW051234300426
44114CB00011B/738